Instructor's M

DEVELOPING CONNECTIONS

Short Readings for Writers

SECOND EDITION

Lorraine Lordi
Judith A. Stanford

Rivier College

Mayfield Publishing Company
Mountain View, California
London • Toronto

International Standard Book Number: 0-7674-1128-5

Manufactured in the United States of America

Mayfield Publishing Company
1280 Villa Street
Mountain View, California 94041

CONTENTS

ANTHOLOGY   26

TEACHING *DEVELOPING CONNECTIONS*

Suggestions in this guide are intended to open possibilities, not to dictate absolute answers or insist that certain patterns or processes must be followed as you teach *Developing Connections*. Both in the text and in this guide, we've tried to include writing topics and suggestions for discussions that will work with diverse teaching and learning styles. We would be most interested to hear from you. Let us know what has worked well for you and your students and what has not. Most of all we would appreciate your sending suggestions for innovative teaching strategies that we might include in the next edition of this guide.

STRATEGIES FOR MAKING CONNECTIONS

In any classroom, but most certainly within a classroom where the subject matter addresses diversity in a lively and often controversial way, instructors are faced with complex, often frustrating tasks, such as

- promoting lively discussion that includes all students and does not degenerate into petty bickering between a few outspoken individuals

- encouraging honest, yet fair-minded writing that takes risks but indicates an ability to communicate clearly and logically

- developing innovative ways of assessing students' writing to encourage both achievement of high standards and growth of self-confidence

Creating a Community in the Classroom

Because students will be sharing their ideas (and probably their writing) with each other, we believe it is essential for them to develop a sense of trust as early in the semester as possible. Developing trust begins with breaking down barriers and challenging the easy assumptions that exist before people

1

get to know one another as individuals. You may want to consider these possibilities:

- Learn your students' names as quickly as possible and use their names whenever you speak to them. This is a painless way to help students learn each other's names. (One way to learn names early in the semester is to arrive at the classroom several minutes before class begins and to ask each student's name as he or she enters. Then jot down a brief descriptive phrase next to the student's name on the roster. Once the student has chosen a seat, write his or her name on a diagram of the room's chairs, and ask students to take the same seats for the next few classes.)

- Arrange the chairs so that students can look at each other as they speak rather than funnel all responses through the instructor. (A circle or semicircle of desk-chairs, for example, or standard chairs grouped around a seminar table.)

- Take a few minutes of the first class to do a get-acquainted exercise. This can be very simple: Ask each student to jot down three statements describing him- or herself and then have students give their names and read their statements. (You may want to prepare your own three statements as well.)

  For a more time-consuming but effective exercise, ask students to prepare to introduce any of their fellow classmates to you. Give the students fifteen or twenty minutes to learn each other's names, and leave the room. When you return, ask each student to introduce the student sitting to his or her left.

Developing an Interactive Classroom

For a course that teaches reading, writing, and critical thinking, we believe an interactive structure works more effectively than a traditional, lecture-dominated structure. In the interactive classroom, students and instructor work together. The instructor does much more than simply prepare a lecture and deliver information to students. Students do much more than read assignments, listen to lectures, and deliver unprocessed information back to the instructor through exams and papers.

In the interactive classroom students take responsibility for their own learning while the instructor provides whatever help and encouragement they need to accomplish this task. Some strategies that encourage interactive learning include the following:

Divide a large class into small groups for discussion or for work on writing projects. Often students who are uncomfortable speaking or asking questions in a large class situation are more at ease when they work in groups of three to five. Consider the following issues:

- *Should students choose their own groups or should the instructor assign groups?* We like to vary the approach, sometimes asking students to form their own groups and sometimes assigning groups to ensure a variety of voices.

- *Should instructors participate with groups or stay away?* Opinion varies greatly among the proponents of group work. Most believe that instructors should interfere very little. Some believe that the instructor should leave the room entirely while groups meet. Another approach calls for the instructor to sit alone for the first few minutes, reading or writing and not looking at students. After the groups have started discussion, the instructor moves around from group to group, mostly listening, but occasionally responding to a question or making a comment.

- *Should groups always or nearly always work toward or reach consensus?* We think it's important to stress that the point of much group work is to discover multiple possibilities. Certainly students should be encouraged to think critically about each other's ideas, but it's not always necessary or desirable that a group arrive at a single, neatly planned response.

- *How should small group discussions be structured?* While there are many ways to structure small groups, here's one possibility:

  Ask students to give their names before they speak for the first time, and encourage them to address each other by name.

  Ask that one student volunteer to be the group recorder, to take notes and speak for the group if the class reconvenes as a whole.

  Encourage students to be sensitive to the dynamics of the group; make each responsible for speaking, without dominating, and for seeking out the opinions of those who may not volunteer their ideas as readily.

Conduct a listening workshop. To stress the importance of listening as a learning device, consider conducting a workshop early in the semester to encourage students to think about listening skills. This is a good way to open the door for strong, productive classroom conversation. A listening workshop may take between twenty and thirty minutes and follows these steps:

1. Explain the process of asking open-ended questions. To begin with, suggest that students ask only questions that begin with the following words: what, where, when, who, and how. (The word *why* tends to set up defensive situations.)

2. Based on the instructions above, invite students to interview you for two minutes (take one question from as many different students as possible during this time).

3. After the time is up, ask students to summarize what they have heard and invite several volunteers to read their summaries aloud.

4. Discuss the differences and similarities in these summaries, stressing strengths rather than weaknesses.

5. Now students should be ready to listen to one another. Ask each student to pair up with someone he or she doesn't know and decide who will "speak" first and who will "listen."

6. The "listener" asks a general, nonthreatening question such as "What is your favorite season?"

7. As soon as the "speaker" answers, the "listener" begins to ask appropriate questions, based on the guidelines given in the first suggestion in this list, for two minutes.

8. When the time is up, the "listener" summarizes what the "speaker" said.

9. The "speaker" then briefly responds to this summary.

10. Now the partners switch roles, and the new "speaker" answers the questions, as the pair repeats steps 6 through 9.

Be aware of the dynamics of full class discussions.

-   The arrangement of chairs can help or hinder free exchange of ideas. Arranging chairs in a circle or semicircle—so that students can see each other as they speak—usually works best.

-   Students often choose to sit in the same seats at each class, and "silent ghettos" (areas of the classroom from which no voices are heard) can develop. You may be able to break the silence barrier by sitting or standing in that area of the classroom and speaking directly to students who are sitting there. If anyone makes eye contact with you, direct your next query or "long pause" toward him or her.

- Some students may dominate discussion. To allow students who may be shy or think more cautiously than others the time to speak, watch the clock and, at some class meetings, announce halfway through the session that you appreciate the hard work and thoughtful observations made by those who have already commented. Then encourage these hard-working talkers to sit back and enjoy listening for the rest of the class period (or for the next ten or twenty minutes or whatever seems right) while those who have not yet spoken offer their ideas. Sometimes you may have to wait many seconds before one of the quieter students volunteers, but once one has spoken, the floodgates open. (This strategy works best if the discussion has started with a warm-up writing. Then everyone has some thoughts committed to paper. Reluctant speakers may be willing to read what they have written as a way to begin their participation.)

- Gender issues can affect classroom participation. Research suggests that men tend to dominate discussions and to interrupt more often than women. In addition, discussion often follows "gender runs." If a man speaks first, other men are likely to follow. When a woman does break into the conversation, others usually follow her. If you notice a gender run going on too long, consider interrupting and directing the discussion to someone of the opposite sex. Watch students carefully and choose someone who looks like he or she is waiting with something to say.

- Cultural differences may also affect class participation. For instance, some silent students may come from cultures where teachers do all the talking and it is considered extremely rude to challenge or question a statement made by an authority figure. Students whose first language is not English may fear that others will laugh at or be impatient with their hesitant or less-than-perfect speech. You may want to meet privately with these students to encourage them to speak and to let them know that you respect their concerns. Consider suggesting that they arrive in class with a prepared observation or comment on the assigned reading. Let them know that you will invite them to offer their prepared observation before you call on other students. This strategy gets them speaking in class but alleviates the pressure and anxiety of not being able to "find the right words"or of not having developed the ability to jump into the middle of a lively discussion.

Require at least one office conference for each student. Seeing students on a one-to-one basis as early in the semester as possible gives them an opportunity to discuss any concerns they may have about the class. The early visit also shows them the way to your office, lets them know that you are available, and encourages them to return with any questions or observations they may have about the assigned reading and writing.

WRITING TO MAKE CONNECTIONS

In addition to interactive discussion, varied approaches to writing lead to a classroom climate that promotes trust, mutual respect, and both intellectual and personal growth. While none of the following suggestions replaces the formal paper, written out of class, they encourage students to see writing as a process that takes many forms and has many different purposes.

Warm-Up Writings

Ask for a few minutes of brainstorming on paper to get students writing early and often. Encourage them to use what they have written to initiate discussions (or to restart a discussion that has ground to a halt or is bogged down in repetitive responses). Usually if you ask them to gather their thoughts through jotting down responses to a topic or a series of topics, students will participate in discussions more readily. Several things happen with such warm-ups: Students discover ideas they didn't know they had; they are forced to really think about the discussion subjects, rather than turning their brains to "rest and recuperate" mode when you ask for a response; and those students who think things through slowly and carefully are given time to ponder and are not left behind while those who speak more readily race from one idea to the next.

The pre-reading questions that precede each selection provide possible topics for warm-up writings. In addition, the photographs and preview quotations that introduce each thematic section offer opportunities for students to ponder and predict the questions, topics, motifs, issues, and conflicts to be addressed in the selections that follow.

Journals

If you plan to have students keep journals during the semester, Section 2 introduces journals and gives you an opportunity to describe what you expect. How many entries do you want students to write each week? Will you specify a length for each entry? Should the entries be carefully written, revised, and edited? Or are the journals to be a place where students can explore ideas, free from the restraints of formal writing?

Journal entries can be the most important writing the students do during the semester; they also give the instructor a fine opportunity to encourage and reward risk-taking. Consider urging students to see their journals as a place to try out ideas and explore responses both to the selections they read and to class discussions.

To provide some guidelines, you might ask students to write at least two entries each week totaling approximately 250 words. At least one entry per week would respond to the selections assigned for reading and discussion during that week. The other entry might explain new thoughts about works read earlier or might comment on issues raised during class discussion. Using these guidelines, then, students would receive full credit as long as they submitted the amount of writing required.

You can use journal entries like warm-up writings—as a way to get discussion started, either by inviting volunteers to read entries they have written on a specific selection or by asking students to read what they have written to each other in small groups. (As always, one member of the group acts as recorder and speaker for the group.)

Journal entries provide a great place for students to explore possibilities for oral responses or for formal papers. To make the journals a safe place— free from concerns about grades—assure students that you will not be correcting or editing what they write in their journals. Consider reading carefully what they have to say and responding with your own comments and questions. As you read students' journals, you might jot observations in the margins and then write a sentence or two at the end of each entry. Keeping these comments as affirming as possible helps students develop confidence in trying out unusual or controversial ideas and approaches to writing. For instance, your comments might praise an intriguing insight, a perceptive observation, or a moving personal connection. Obviously you will not admire or agree with everything students write in their journals. Sometimes entries are boring, shallow, or even offensive. Yet suspending harsh judgment and instead asking questions or suggesting possibilities often leads a halfhearted or hostile student to engage in the class more positively and fully.

Journals provide a way to communicate with students each week in a private and safe setting. For instance, if you notice that a student seems uncomfortable during group work or during discussion, you might make a note in the journal, asking if you can be of help and suggesting that the student come to office hours or talk with you after class. You will often find that students let you know in journal entries about some aspect of the course about which they are either troubled or pleased.

As a structure, you might collect journals once a week and return them on the same day of the following week. To facilitate the handling of journals, ask students to write on loose-leaf paper and submit just one week's entries in a pocket folder. When you return the journals, students remove the entries on which you have commented and store them in a three-ring binder. Then they place their new entries in the folders and hand them back to you. This process means that you carry home slim folders rather than heavy notebooks of varying sizes and shapes.

If collecting and writing responses to journals doesn't fit your course plans, consider the following possibilities:

- Do not collect journals, but ask students to bring them to class to use as a basis for discussion.

- Collect journals but return them with only a check mark to indicate "accepted" or a minus sign to indicate "needs improvement before I can accept it."

- Collect and grade journals, evaluating them in the same way you would evaluate an essay or research paper.

Journal work has many helpful side effects. For example, when students have to write on the assigned readings, they come to class prepared, having thought about what they've read. When you assign reading journals, quizzes are usually neither necessary nor useful. Because students are doing so much writing, they begin to feel more at ease with the process. Journals convince them that writing is a way of thinking, and many of them begin to be more comfortable with formal writing as well. Finally, journals provide a window into thoughts, hopes, and feelings that students might not readily express aloud in class. You have a chance to see what is really going on for them as they read the assigned selections, and often these insights can help you shape future class plans.

In addition to topics for journals that you generate or that your students come up with on their own, each reading selection in *Developing Connections* is preceded by several pre-reading and journal-writing suggestions.

Collaborative Projects

Many of the extended writing suggestions that follow each selection work well as collaborative projects. Students might begin collaborations during small group meetings in class. During their first meeting as a group, encourage students to exchange names and telephone numbers to facilitate arranging further meetings.

When the projects are completed, groups might serve as panels, with each member presenting part of the group's findings to the class. To accompany this presentation, you may ask for one written report from the entire group (a paper that would require a great deal of cooperation and coordination), or you might ask each member of the group to submit a one- or two-page summary of the oral report offered during the panel discussion.

In-Class Essay Series

While organizing this project takes some time, it ultimately promotes a great deal of writing while keeping the instructor's correcting time to a minimum. During the term, have students write a series of essays—perhaps three or four—in class, taking about thirty minutes for each. These writing sessions may be announced or unannounced. After each session, collect and alphabetize the essays by student last name and then store them in a large envelope or folder.

After all in-class essays have been completed, collected, and stored (but not corrected), redistribute them to students during another class period. Ask students to meet in small groups or pairs to read each other's essays. With the help of peers, each student decides which of the three essays to rewrite and submit for a grade. If you ask them to submit the original essay along with the final copy, you'll have a chance to observe and comment on their revision process.

Suggestions that follow each reading selection provide possibilities for the in-class writing series.

ASSESSING STUDENTS' WRITING: DEALING WITH THE SCARLET LETTER

The ways and means of assessing student writing are nearly as varied as individual instructors themselves. As a result, students are often under-standably confused about what good writing really is and may even note that the same piece receives a high grade from one instructor and a low grade from another. Consider, then, the following objectives:

1. Students should understand what you expect in any assignment. Class handouts or class discussion should focus on the requirements for all stages of writing (see examples A–E, pp. 14–18).

2. Established criteria should be appropriately balanced. For example, the content of a piece of writing should be at least as valuable as the surface correctness of a piece (see example C, p. 16). (For most

assignments, most instructors will see content as considerably more important than surface correctness.)

3. Either verbal or written comments (or both) should complement the grade on a piece of writing. Such comments should aim to show students their strengths in writing as well as their specific weaknesses.

4. Students should be allowed to revise initial drafts.

Based on these objectives, consider the following assessment strategies related to two elements of the writing process: revision and final products.

First-Draft Assessment Options

The ungraded first draft. In this case, first drafts do not receive a letter grade. Instead, the instructor either holds a one-on-one conference or writes marginal comments to direct students to areas that deserve praise or require more work. We also know an instructor who tapes her comments while she's reading her students' drafts, and she's found much success with this method of feedback. (Students purchase a blank audio tape as part of their "books and supplies" requirement; they submit their writing in a pocket folder along with the tape; the few students who have no tape player can use one at the library or media center.)

To avoid overwhelming the student (and, just as important, to avoid an overwhelming amount of instructor time), consider limiting comments and directions in order of importance:

High-order concerns: content, focus, organization, development
Middle-order concerns: coherence, style, word choice
Lower-order concerns: usage, mechanics, grammar

Scaling the first draft. We've found that many students want to know where their writing falls within a certain range of writing expectations. In addition to writing comments on drafts, our fellow instructors have found success with marking the drafts with one of the following possibilities:

✔+	✔	✔–
• check plus	check	check minus
(outstanding)	(acceptable)	(needs work)

• Publishable, revisable, rewritable (see example D, p. 17)

• An "If this were a finished product" grade (not figured into a student's final grade)

- A 1/3 draft grade, 2/3 final product ratio

- A "process" grade that measures the apparent effort in composing the first draft

Marking errors. Whatever approach you choose to use, we strongly recommend that you do not correct the surface errors on student papers. By correcting such errors, you end up doing more work than the student. In addition, the student often fails to recognize his or her own pattern of errors. We've found the following options make students more responsible for their own editing issues:

- Put a check in the left margin next to any sentence that contains an error in grammar or mechanics. In the beginning, we've found it helpful to give the student a hint at the error:

    P = punctuation; SP = spelling; a plain check means anything else.

- If a paper has one or two recurring errors, point them out on half the paper and ask the student to locate the rest.

- Correct one error and give a brief explanation of what is causing the error. ("This fragment is caused by an '-ing' word. It needs an auxiliary ['helping'] verb to make the thought complete.") Ask students to locate and correct any other errors of this type.

- Ask students to keep a record of the kinds of errors in their writing. Have them contract to rid their next paper of at least one of these kinds of errors.

Final Products

If you've commented on students' drafts (either through written comments, office conferences, or class workshops), you can usually read and assess final papers rather quickly. At this stage you want to commend the student for any improvements in revising and give the student some helpful advice for future writing assignments.

In addition to short, clear comments, many instructors find an evaluation sheet especially useful in highlighting the strengths and weaknesses in a piece of writing. More importantly, students can immediately see the areas that deserve more attention the next time around (see examples A–E, pp. 14–18)

A Plug for Portfolios

In recent years, many instructors have switched from grading single papers to requiring students to submit a semester portfolio for a final grade. As in the area of assessment, there is no one way to use or grade portfolios in any classroom. However, one thing is certain: Many instructors who have switched to portfolio evaluation swear that they will use this method for a long time to come. The feedback from students has also been positive.

Here are three sample options for a final portfolio's contents:

- Twenty typed pages, including (at the instructor's discretion) formal essays as well as journal entries, creative pieces, reports requiring research, and so on

- Four major papers (three to five pages each) written for the class, including the last formal writing assignment

- A minimum of thirty pages, including reading responses, art work, other class work, and choices from writing assignments

Final assessments on portfolios also vary. Some instructors prefer to assign just one grade to the entire portfolio. Still others assign different grades to specific criteria, such as amount of writing, growth in writing, final products, writing process, the selection process, and the self-reflection process.

Whatever criteria you choose, keep in mind that collecting portfolios at midterm is a good idea, especially for first-year students. That way, students aren't scrambling the last week of school to throw together a collection of their best writing. In addition, a sharing of portfolio entries three or four times over the course of the semester allows students to publish the pieces they most want to share.

Finally, many instructors have found that allowing students to take part in the assessment process (this applies at any stage of writing) makes them more apt to recognize their own responsibility for the final grade their writing earns. Consider questions like these to encourage students to think about what they have written:

- If you had more time, what would you do with this writing?

- What do you feel is the strongest part of this piece?

- What areas might need more work at this point?

- What were some of the problems you encountered while working on this piece?

- What would you do differently the next time around?

- Assess your level of dedication to this writing: poor, so-so, average, zealous, and so on!

- Assign this paper the grade you feel it merits.

Example A

INTRODUCTION TO COLLEGE WRITING—
EVALUATION, PAPER #2: REPORTING

Student: _____   Title: _____

	weak			strong
Focus:	( )	( )	( )	( ).

1. Are the ideas in this paper clearly stated?
2. Is the thesis clear and narrow?
3. Does the writer cover one incident or event?

Development:	( )	( )	( )	( )

4. Does the writer explore the subject in depth?
5. Is there enough information contained in each paragraph?
6. Is there any unnecessary information?
7. Does each paragraph contain just one topic?

Organization:	( )	( )	( )	( )

8. Is the information in this paper organized in a logical, clear manner?
9. Is the introduction interesting and informative?
10. Does the paper conclude effectively?

Details:	( )	( )	( )	( )

11. Does the writer provide good, concrete details?
12. Does the writer use strong, active verbs?
13. Does the writer use clear, descriptive language throughout?

Mechanics:	( )	( )	( )	( )

14. Punctuation
    Fragments
    Spelling
    Agreement

Overall Grade/Comments:

Example B

THE WRITING PROCESS EVALUATION

ESSAY: _____   Student Writer: _____

Category	Maximum Points	Your Points
FOCUS: This essay has a clear idea or purpose; the introduction indicates the topic and direction of the essay; body paragraphs support the theses.	20	
ORGANIZATION: Ideas flow in a logical, clear manner; transitions are effective; conclusion relates to introduction and fulfills the reader's expectations .	20	
DEVELOPMENT: The subject is explored in depth; specific facts and details add vitality to the essay; all information is necessary.	30	
STYLE: The writer's manner is suited to the audience and purpose; words are precise; writing is strong and clear; sentences are constructed to enhance focus and add vitality to the essay.	15	
MECHANICS/EDITING: Sentences are properly constructed; punctuation is correct; subject, verb, and pronouns agree; the writer has proofread for spelling and typographical errors.	15	
OVERALL SCORE	100	

SUGGESTIONS FOR REVISION/FUTURE WRITING ASSIGNMENTS:

Example C

BASIC WRITING

ESSAY: _____   **Student Writer:** _____

	Maximum Points	Your Points
FOCUS: Paper deals with one topic or idea. Thesis is clear and controls subsequent paragraphs.	20	_____
ORGANIZATION: Ideas follow in a logical, clear order.	20	_____
DEVELOPMENT: Statements are supported through examples, vivid details, and facts.	20	_____
INTRODUCTION AND CONCLUSION: Opening paragraph captures the reader's interest and indicates the author's purpose in writing; conclusion sufficiently wraps up information and leaves the reader with something to think about.	15	_____

GRAMMAR AND MECHANICS

	Number of Errors
Sentence fragments	_____
Run-on sentences	_____
Subject-verb agreement	_____
Verb tenses	_____
Quotation punctuation	_____
Spelling errors	_____
	25	_____

| **Total:** | 100 | _____ |

GENERAL COMMENTS/SUGGESTIONS FOR REVISION:

Example D

COLLEGE WRITING I: EVALUATION

Student: _____

WRITING IN PROGRESS/DATE: _____

PUBLISHABLE: The writing at this point has a clear and perceptive point. It is also well supported and developed with reasoning and details. The organization is effective and appropriate, as is the style of writing in this piece. Only minor spelling and mechanical errors still remain at this point.

Suggestions:

REVISABLE: The idea behind this writing is good and shows considerable evidence of planning, but at this point, the writing needs more attention to one or more of the following areas: organization, adequate support and development, smooth transitions between major points, style, or mechanics.

Suggestions:

REWRITE: The writing is unacceptable college writing at this point because of one of the following:

1. The writing makes no consistent or useful point. No clear purpose is evident.

2. The writing is poorly constructed and was hastily done; the reader has trouble following the main idea.

3. The writing contains so many sentence-structure errors that the message is troublesome to decipher.

4. The writing does not fulfill the assignment.

Suggestions:

Example E

COLLEGE COMPOSITION: SUGGESTIONS FOR REVISIONS

Student Writer: _____

TOPIC: _____

CLARITY: Thesis statement, introduction

INFORMATION: Development, support, organization

APPROPRIATE LANGUAGE: Tone, correctness

PLANNING THE SYLLABUS

The structure of *Developing Connections* allows great flexibility as you design your syllabus and plan for each class. Here are descriptions of the book's features and suggestions for ways to use them.

Section 1: Critical Reading and Thinking: Recognizing Cultural Contexts

Whatever direction your course takes, this chapter introduces the concept of cultural contexts and practical reading and thinking skills. Students also begin writing with short, paragraph-level suggestions. See page 22 for detailed teaching suggestions.

Section 2: Reading to Respond

Exercises and explanations lead students through the process of responding to what they read by making marginal notes and writing journal entries. Writing suggestions elicit from students paragraphs expressing responses to readings.

Section 3: Reading to Understand

To move from response to critical thinking, students learn how to identify unfamiliar words, summarize, identify main and supporting ideas, and make inferences. Writing suggestions elicit paragraphs explaining the inferences students make as they read.

Section 4: Reading to Evaluate

After explaining response, summary, inference, and evaluation, this section shows students how to develop criteria for evaluating what they read and leads them through the process of applying those criteria. Exercises show how discussion can contribute to the process of evaluation, and writing suggestions elicit evaluations students have formed while thinking about sample readings.

Section 5: A Writing Process

This section moves students from paragraph-level writing to longer papers. The step-by-step example of a student's paper demonstrates one approach to developing a writing process.

Photographs

Each thematic unit begins with a series of photographs that relate in some way to that unit's theme. You'll find specific suggestions for students to write about and discuss these photographs included in the text.

As a general teaching strategy, you may want to use these pictures to encourage students to develop the pre-reading strategies of prediction and early response. They might write journal entries or brief warm-up commentaries in class suggesting what expectations, questions, issues, and concerns they'll have in mind as they read the selections you've assigned.

Pre-Reading and Journal-Writing Suggestions

These suggestions precede each reading selection and serve as bridges from the students' own experiences and current knowledge to the ideas, feelings, and information the author offers. Like the photographs, these suggestions provide opportunities for students to sharpen their predicting skills and will arouse interest in the reading that follows.

Reading Selections (Anthology)

Each of the seven thematic units in the Anthology provides a variety of selections that reflect a particular theme related to cultural contexts and exemplify the motive for writing explained in the introduction of the section where that theme appears.

Nonfiction reading selections include essays, newspaper articles, journals, reviews, letters, and excerpts from books. You may want to point out these different approaches to writing and ask students to experiment with various approaches in their journals or assigned papers (if you consider these approaches useful for the writing goals of your course).

In addition to the nonfiction readings, each thematic unit offers a short story or a poem. Reading these examples of imaginative literature encourages students to think in different ways about the themes they are pursuing and perhaps to try some imaginative writing of their own.

Suggestions for Writing and Discussion

Following each selection are suggestions you can use for discussion or for the basis of a writing assignment. Many of these topics encourage students to pursue the aim exemplified by the selection they have just read. If you prefer to have students work on one writing aim at a time, you might reserve those topics that suggest an aim different from that section's focus for discussion or for informal writing, such as journal entries or in-class warm-ups.

Suggestions for Extended Thinking and Writing

These suggestions, which follow each selection, work well as group projects. Many provide the option of oral reports, panel discussions, or debates to accompany, or even replace, written response.

Of course, many of these topics work equally well as individual assignments, and if you ask for research as part of the course work, you'll find plenty of possibilities here.

Suggestions for Making Connections

At the end of each thematic unit are writing and discussion suggestions that encourage students to see relationships among the selections they've read. Like the Suggestions for Extended Thinking and Writing, these suggestions offer many opportunities for group reports, panel discussions, and debates, as well as for individual writing projects.

To provide additional reading selections for these topics, consult the cross-reference list for each reading (see, for example, p. 26).

Rhetorical Index

If, in addition to stressing the aims of writing, you also want to acquaint students with the rhetorical modes, the text offers a rhetorical listing of selections following the table of contents.

Overview: Sections 1–5

Sections 1–4 each introduce approaches to reading, writing, and thinking. Examples, explanations, and exercises lead students through processes essential for success in their other college courses, as well as in their future professional lives. Writing assignments in these four sections are most appropriate for short, paragraph-length pieces. While these four sections are most often taught in sequential order at the beginning of the semester, some instructors like to teach one chapter, then go on to working with a section in the thematic anthology, return to another reading-writing section, and so on.

Section 5 provides a step-by-step process for writing. This section, which leads to writing fully developed essays (three to five pages or longer), is most often taught after the first four sections. Some instructors, however, prefer to teach this chapter first and thus encourage students to write longer essays from the beginning of the semester. The first four chapters might then be interspersed throughout the semester to introduce or reinforce skills students are using as they work with readings in the thematic anthology.

Section 1: Critical Reading and Thinking: Recognizing Cultural Contexts (text pages 3–10)

The explanations, examples, readings, and exercises in this section provide the rhetoric of cultural contexts that will serve as a basis for reading, writing and thinking about, and discussing the selections in the rest of the text. Depending on the abilities and interests of your students, you will probably want to spend about one week working through the issues addressed by this section.

Students in some classes will easily identify themselves as coming from various cultural contexts; perhaps the class will include several international students or students from very different racial and ethnic groups within the United States. On the other hand, students in some classes may look around and evaluate their peers as all belonging to the same culture. Whatever the makeup of the class, you may want to encourage students to broaden their definition of culture (as suggested by "What Is Culture?" on text p. 5). When students recognize that any given group of human beings is diverse in many ways and that they themselves are members of several culture groups, they often feel more open to studying and thinking about the concept of cultural context.

Because issues of diversity have had great play in the media (and in the political forums of the 1990s), students may have strong opinions about looking at and paying attention to the astonishing variety of cultures within their college, their town or city, their state, and their country. To begin discussing these issues, ask students to write and talk about their responses to Horace Miner's "Body Ritual among the Nacirema" (text p. 3). Most students enjoy Miner's piece, and many, with a little urging, will create an "anthropological report" on an aspect of their own lives. The laughter that accompanies writing and reading the misconceptions and skewed interpretations contained in these reports often leads to a willing suspension of the suspicion and wariness some students feel when they hear the word culture.

Developing Connections tries to go beyond offering a collection of readings that reflect America's diverse society. As the opening section suggests, most students have been exposed to the idea of cultural contexts, yet many regard this concept as just another remote philosophical vision visited on them by their teachers and textbooks. "Why Read and Think in Cultural Contexts?" (text p. 6) suggests many reasons for thinking critically in the context of different cultures. To encourage discussion of these reasons, ask students to do one or both of the two exercises related to the journal entry by John Coleman (text p. 8).

Section 2: Reading to Respond (text pages 11–21)

To practice, students may first read "Reading to Respond" (text p. 11) and then complete the exercise relating to Gloria Bonilla's "Leaving El Salvador" (text p. 15). Students might share their marginal responses and the resulting journal entries in small groups, noting similarities and differences in their observations. You may want to point out that they need not try to reach consensus. Multiple readers add new possibilities, and the group's strategy should be to open options, not to enforce a single view. When they finish discussion and convene as a class, you might ask them to write for two or three minutes on their reaction to the group discussion. Were they surprised by any of the observations they heard? Did they discover any new ways of looking at the selection? Did they change their original response in any way?

Section 3: Reading to Understand (text pages 22–30)

After thinking about and discussing their initial responses, readers need to delay the rush to judgment by returning to the text and discovering exactly what the writer is saying. This step is particularly important when the topic is controversial or when it is offered from an unfamiliar perspective, both of which are often true of writings with multicultural themes. To work on these skills, students may begin by reading "Understanding Unfamiliar Words" (text p. 22) and doing Exercise 1 (text p. 24), which works well as a small group task. After reading "Summarizing to Understand the Main Idea and Supporting Ideas," students may do Exercise 2 (text p. 27). To continue the process of reading to understand, Exercise 3 (text p. 28) provides further practice in summarizing. You might ask students to summarize one specific essay, rather than giving them the options provided by the assignment in the text. If all students summarize one essay, you could collect them, make transparencies for the overhead projector from some, and use them as the basis for a class on summary writing.

Understanding inference is extremely difficult for many students, yet failing to understand connotation or misunderstanding shades of meaning can lead to serious breakdown in communication and comprehension. To assess your students' facility with making inferences, ask them to read "Reading to Understand Inferences" (text p. 29) and to do Exercise 4 (text p. 30) as an in-class writing exercise to be discussed with the class.

Section 4: Reading to Evaluate (text pages 31–39)

Most college courses and most professions require making evaluations, yet many students do not understand the difference between summarizing the author's ideas and making a judgment of their own. "Reading to Evaluate" (text p. 31) takes them through this essential process step by step, and the exercise (text p. 38) provides a way to practice the strategies they have just read about.

Section 5: A Writing Process (text pages 40–58)

While the step-by-step explanations and examples in this section should be extremely valuable to students, it is one of the most challenging sections to teach. Because of the demonstration approach, students who read this section carefully usually have few questions. Many students, however, simply skim through and fail to digest the process that is examined.

We have found it helpful to assign the section very early in the semester and to ask students to walk through the various steps as a class exercise while they work on their first papers. For example, we frequently ask them to work in groups on discovery strategies and then return to the full class to report on and discuss what happened in the groups. We include in the group work a request that as they complete their work they compare their experiences with the experiences of the student described in the book.

Another possibility is to ask that students include a process commentary with their own first paper. While not nearly as long or complex as Section 5 in the text, this commentary should describe and explain the steps the student took (obviously, not exactly the same as the steps in the book) while completing the paper.

Still another way to use this section is to ask students to bring the text with them for their first office conference. If they express (or you have noted) particular difficulty with any stage of the writing process, walk them through the explanation in the text to see if they understand the options presented. We've discovered that often students have barely looked at the text and that gentle and nonjudgmental direction toward the help offered there is often gratefully received.

ANTHOLOGY

ARRIVALS, ROOTS, AND MEMORIES

Photographs

1. What kinds of arrivals, roots, and memories are suggested by these two photographs? Explain, for example, where you think the people in these pictures are coming from and where you think they are arriving. What kinds of memories might be going through their minds at this time?

2. Write a journal entry from the point of view of one of the people in this photo. As you write, keep in mind the theme "Arrivals, Roots, and Memories." Write a brief paragraph, giving the biographical details you have assigned to this person (age, occupation, nationality, and so on).

Mary Antin, The Promised Land (text p. 62)

Suggested Cross-Reference Readings	Text Pages
Letter to My Mother	106
The Teacher Who Changed My Life	186
Finishing School	179
A Brother's Dreams	221

Before reading and discussing this piece, consider inviting students to share their responses to one of the journal-writing suggestions. To foster listening skills, ask other students to listen for rich details: images, sounds, senses, colors. After students respond positively to the shared journal writings, you may even push them one step further by asking them to become curious listeners. What more would they like to know, especially in the way of details and images? This interviewing step is a subtle move toward revision strategies, which you may want to point out as a way of concluding this part of the discussion.

Once students have recognized the importance of details in their own writings, remind them to take note of the details that Antin uses in this piece. Such specifics should be used to support responses to the first discussion question. Answers to this first writing and discussion (WD) question will vary, but most will probably mention the author's initial feelings of excitement, gratitude, awe, and understandable nervousness. You may want them to point to exact places in the text where Antin shows gratitude or other feelings.

Once students have discussed the character's reactions to a new place, in question 1 (WD), ask them to compare their own journal responses with the character's experience. Did they have the same feelings in a new place? In what ways did they respond differently? Such a simple prompting shows students that their own experience is invaluable in understanding a text.

When they come to questions 2, 3, 4 and 5, students should feel comfortable combining their own experiences and their prior knowledge with the details provided in the text. For example, question 2 asks them to evaluate their attitudes toward education with the author's. For students to recognize their own attitudes, they'll need to brainstorm memories for specific examples; to verify the author's attitude toward school, they should provide representative details and incidents from the text. (If you have selected "The Teacher Who Changed My Life" as a paired reading for this piece, students could also compare Gage's reaction to education: the system, the work, the students, the teacher.)

To expose students to the structure of comparison, you may want to list two or three different experiences on the board and ask students to come up with categories that come under the general topic of "education." Selecting points such as "teachers," "students," "subjects," and "workload" will give them a start toward recognizing the different aspects a reader can choose when comparing subjects.

Questions 3 and 5 continue with this comparative mode of thinking. Students are asked to compare the freedoms in other countries with those in the United States, and to compare Antin's family with modern American families as portrayed on television and in movies.

If you have chosen "Finishing School" as a paired reading, question 4 will undoubtedly elicit different responses on the topic of names. How important are they? What do they reveal about us? Why would one individual so willingly give up her name while another clings to hers? To offer students an intriguing challenge, ask them to dwell not on the differences between the two female characters but on their similarities.

Again, by discussing this question first in small peer groups, students may see that a name has the power to elevate or destroy us. It is also very often a mirror of who we are and what we value most.

The Suggestions for Extended Thinking and Writing give students the chance to do the following: write a comparison paper based on their own experiences (suggestion 1); write a process description essay (suggestion 2); and write an objective essay that depends on questioning strategies and listening skills.

Toshio Mori, The Woman Who Makes Swell Doughnuts (text p. 68)

Suggested Cross-Reference Readings	*Text Pages*
I Show a Child What Is Possible	170
Going Home	225
How Boys Become Men	260

According to veteran writing teachers, if you want all of your students to write with energy and ease, just ask them to write about food! Students are guaranteed to enjoy writing and sharing their journal freewrites on food. Pre-reading suggestions 1 and 3 allow them to do just that.

Most students will see the connection between the doughnuts and the author's relationship with "Mama." At times the author writes so abstractly that students should be justified in wondering, "What the heck is he saying?" The first paragraph is a prime example, especially when the author writes about the dance and the "stirring of the earth."

For most of the essay, however, the author speaks clearly and directly, and students should be able to identify with the caregiving qualities of an older person as well as the subsequent comfort and love felt by the younger person (suggestions 2 and 4). Students may even see similarities between "Mama" and the older person they wrote about in their journals.

Comparing the experience of eating a doughnut to "an event of necessity and growth" may be a bit confusing (here again the author is speaking in a more poetic voice). Nonetheless, students should be able to understand how love and compassion help another person grow (suggestion 3).

In order to set up question 5, you may want to ask the students to list how many people in their lives they can sit with in silence and still feel comfortable. A relationship that is comfortable in silence, then, is a close and deep one—one that has overcome or transcended the limitations of awkwardness. Neither person feels the need to entertain, inform, or question the other; each can accept the other's company alone as sufficient, fulfilling, and comfortable. No words are needed.

Before students discuss the last question, you might ask them to look back to their journal entries on food. Besides the food itself, what else did students find themselves writing about? Chances are that along with the food, students conjured up memories of past times and people. Thus, the doughnuts, like the students' own food choices, keep alive memories of special people, special times, and special relationships.

If you want to push students beyond the literal, ask them in what ways a doughnut might be especially significant or symbolic. (The complete circle, like a ring, can be a symbol of lasting love, loyalty, and contentment.) You could also encourage them to evaluate the significance of their own food choices from their journal entries.

After completing this selection, students might write a descriptive essay on an older person. Their initial journal entry would be a great springboard for writing here.

Joe Klein, The Education of Berenice Belizaire (text p. 72)

Suggested Cross-Reference Readings	Text Pages
I Show a Child What Is Possible	170
"I Just Wanna Be Average"	193
Surfing's Up and Grades Are Down	305

How and why do students learn? These are two good questions that will spur endless discussion and thinking as students delve into this essay. Before actually getting into the student's situation in this piece, consider first discussing students' pre-reading responses: What do most Americans think about education? Do they see education as a right, a privilege, or something else? Instructors might even set up one-on-one interviews or small group discussions that center around these ideas. Students should be encouraged to speak honestly and openly about their own experiences as well as their observations of fellow students.

After this initial discussion, students can think about the specific qualities and circumstances that were the causes of Berenice's success. A triangle of sorts should emerge as a result of this discussion: family, school, and the student are each pivotal points in learning. Students may recognize other factors as well, such as cultural values and expectations and personal circumstances that create a need for learning. Once these causes are uncovered, consider encouraging students to identify which factor they consider most important. Of course, there is no one right answer here. But after reading this piece, most students will recognize that, despite her situation, Berenice not only survived but thrived as a learner.

How did that happen? While many factors matter—the mother's encouragement, the teachers' recognition of Berenice's talents, her own natural bent for remembering—Berenice herself points to a factor in learning that is often overlooked. In paragraph 4, she says that hard work was the key to her success. It's not a glamorous or magical answer, but it was the crucial factor in this student's success.

Is hard work a characteristic that the American culture fosters? As you ask students what they think, you might invite them to offer concrete examples and to base their conclusions on specifics that they have observed or experienced themselves. What many might find, as the author himself infers in the last line, is that some American students—and perhaps a growing number of Americans in general—do not see the importance or value of something as mundane as hard work.

Miguel Torres, Crossing the Border (text p. 76)

Suggested Cross-Reference Readings	*Text Pages*
Letter to My Mother	106
"I Just Wanna Be Average"	193
A Brother's Dreams	221
The Teacher Who Changed My Life	186

To help students appreciate the circumstances under which many immigrants come to America, it is important that they write openly and honestly on pre-reading question 1. Describing an idyllic country may be helpful, as well (question 2).

Instructors who have used this piece with their students report that whether or not students like this essay, they all seem to have strong opinions about the decisions Miguel makes. Although a few students are sympathetic to Miguel's plight, many see him as a freeloader who is just out to make fast money and have a good time (question 1).

Allow students to air their opinions without too much commentary on your part. Miguel's plight has a remarkable tendency to surface again and again over the course of the semester!

Perhaps when students stop to think about the connotations of the word coyote, they will see that Miguel might be partly a victim of his circumstances. (Coyotes are wild animals who are usually loners, preying on every and anything to keep alive—even the remains of dead animals. In Native American cultures the coyote is often considered clever and tricky.) Miguel may be a victim of the myth of the American Dream as well.

As the first paragraph shows us, Miguel is probably functionally illiterate in his own country and has lived in poverty, probably without the guidance and support of a father (question 3).

Therefore, some students might give him credit for being a survivor, someone who doesn't give up. Some may also see his laid-back attitude as a character strength. It is most likely, however, that students will note Miguel's lack of ambition, especially as far as education goes. Perhaps most students would not choose Miguel as a friend; however, intriguing responses result when they turn the question around and ask themselves if they have any friends who are similar to Miguel (suggestion 4).

Each time Miguel is shipped back to Mexico, he seems to become more comfortable and efficient in entering America illegally. He seems to fear less and know more about the system. Considering these points, it seems probable that Miguel will eventually find a way to stay in this country (suggestion 5).

Although most students don't believe Miguel should be living in America, they will agree that life for him in America is better than life in Mexico. Predictions should vary for where Miguel will be in fifteen years. Some students may see him back in Mexico parking cars; others may see him in America, married to an American and working as a farm laborer or on welfare; a few may envision his getting an education and becoming a member of the legal American workforce.

Students will be divided about whether Miguel's presence in America is a hindrance or a contribution. Although this question begins with Miguel, it usually ends up in the area of immigration in general. Many students see immigration as an ever-increasing problem in the United States, while others see immigrants as a rich resource for all of us (suggestion 8).

One suggestion for extended thinking asks students to write an imaginative piece from an immigration officer's point of view. If you want students to work in a narrative form, suggestion 2 allows them the chance to use their own personal experience as a way to sympathize with Miguel's plight in some way.

John Tarkov, Fitting In (text p. 81)

Suggested Cross-Reference Readings	Text Pages
From Father, with Love	109
The Men We Carry in Our Minds . . . and How They Differ from the Real Lives of Most Men	286
Modern Courtesy	310

This is a touching, beautiful essay—rich in detail, voice, and truth. Most students enjoy this selection, not only for its subject matter but also for the author's style: clear, direct, and deceptively simple. Before beginning this piece, consider discussing the differences between parents and children. Students could share their initial pre-reading responses, either in small groups or one on one. Also, any students whose parents have come from another country might be encouraged to explain the differences between themselves and their parents as they adjust to life in a new country.

Students should also be encouraged to annotate this piece, indicating the phrases, words, and details that strike them the most. They should also begin to raise questions that will lead them to a deeper understanding of this relationship: Why does Tarkov include details about the baseball and glove (paragraph 2)? Why the details about the father's death (paragraphs 4–6)? Why the details about the kind of food the father and son share and the programs they watch together (paragraph 9)? Questions such as these lead students to become close readers and deeper thinkers. Not only do the details serve as visuals for the reader, but to the close reader, they serve as connections to deeper issues—issues of relationships, guilt, sorrow, regret—that all humans experience.

Patricia Hampl, Parish Streets (text p. 85)

Suggested Cross-Reference Readings	*Text Pages*
Letter to My Mother	106
From Father, with Love	109
The Loudest Voice	211

All three pre-reading questions ask students to think about religion. You may ask students to write on all three and then choose one to share, or you may ask that students pick one question on which to write. Either way, leave some time to discuss childhood memories of religious influences, as religion (or rebellion against religion) is undeniably one of the more powerful cultural influences for most of us.

Encourage students to read this piece straight through, just for enjoyment. You may want to ask what they thought of Hampl's story. Did they like it? Were they bored by it? Could they relate to it in any way? How?

The categories given in question 1 can be used to promote individual writing. They would also work well as a small group activity. For example, divide the class into five groups, each group taking responsibility for one topic. Then allow time for the group to find details and come to some conclusions about what the author's point might be on each specific topic.

After going through each category, students should have no problem generalizing about Hampl's childhood. Adjectives they might use include simple, curious, traditional, ritualistic, wonder-filled, sensitive, fun, funny, and Catholic.

For question 3 you might ask students to come up with a word or phrase other than crazy to explain what Hampl means. Terms such as ridiculous, nonsensical, illogical, and stupid may be a part of their listing here.

Mr. Bertram (question 4) is an interesting character, and he's in this essay for a most important reason. If you allow several students to voice their opinions on this question, some may see that Mr. Bertram is an obvious contrast to the ritualist Christians in this piece. You may push students to recall if they know anyone like him. What exactly does Mr. Bertram believe in? Is he an unbeliever? Or is he a spiritual man? (Discussing the difference between religious and spiritual often evokes thoughtful responses.)

Students should refer to the text itself as they respond to these questions. Certainly, Mr. Bertram is a man who wants to believe. (He looks to the sky twice.) He is also a very private, contemplative person who takes little stock in conventions established by his fellow humans. (He looks "resigned as Buddha" among his dandelions, and he admits that "God isn't the problem" with the world.)

Jimmy Guiliani is also a relatively complex character. He is the rebellious kid on the outside, the one who doesn't do things the proper way. Unlike Mr. Bertram, he lashes out at people and does hurtful things. While one says to the narrator, "You can't know," the other says, "Try to see things from my point of view" (question 4).

The last character we meet is the rather plain parish lady. Although she is one who goes unnoticed, Hampl remembers the woman smiling at her. This woman is a symbol of deep faith, of peace and contentment within religious confines. She provides an example of people to whom religion can bring understanding and contentment, and she completes the trio of characters who have found their own ways of coping with the world. One draws away from it (Mr. Bertram), one lashes out at it (Jimmy G.), and one (the parish lady) functions quietly within it. You might ask students to speculate whether Hampl seems to prefer one of these approaches over the others. Perhaps by describing the parish lady last, the author grants her a nod of approval (question 5).

Most students will be able to figure out the phobia part of xenophobia. If not, have students list other phobia words, until they realize that this word has something to do with fear. If no one knows the term, assign students to look it up or tell them that this word means "a fear or hatred of foreigners." Therefore, religious groups have a way of causing people to be afraid of those who are outside of their beliefs, their own church.

Clearly Hampl does not promote religious xenophobia. Most of the examples she uses show the irrationality of such a religious attitude—the "us against them" frame of mind that so often crops up between peoples of different faiths.

As a closure to this piece, you may want to discuss any present-day tensions that are a direct result of the tenets held by different churches. Is the Catholic Church still this rigid today? What churches tend to be most "open" to people of other faiths? Why, you might ask, do discussions about religion tend to end up as shouting matches instead of educational ventures?

The Suggestions for Extended Thinking and Writing continue this theme of religion and its influence on the individual. Students may choose to explore the benefits of being a member of a church community (suggestion 1), they may choose to write imaginatively on any one of the characters mentioned in this piece (suggestion 2), or they may opt to use their prior experiences as major support for an extended essay (suggestions 3 and 4).

Isaac Bashevis Singer, The Son from America (text p. 95)

Suggested Cross-Reference Readings	Text Pages
Letter to My Mother	106
The Teacher Who Changed My Life	186
What Is Intelligence, Anyway?	176

Singer is recognized not only as a great writer of fiction but also as a great philosopher. His stories often work as allegories, as parables, or, as students might better recognize, as fables. The central theme in this story will make students think and question on several different levels: Singer's characters lead readers to ponder the difference between a "need" and a "want." In considering this question, students must define and contrast to formulate an answer. Further, their thinking can take on an analytical nature: What circumstances shape one's needs and desires? Which one of these forces is more powerful? Why? Last, students can synthesize their own experiences and observations with the facts and events in the story.

To stress the importance of close reading and drawing conclusions based on details, students might be encouraged to read this piece slowly, noting the details along the way. For example, they might look for clues that tell them what this little village looks like. Students who are drawn to art may be willing to sketch their image of the town for the class. Students could also draw pictures of Berl and his wife, considering all the details listed in this piece. And the son—what does he look like? For students who are not inclined to sketch, a search through travel magazines might provide materials for a collage representing the village and its inhabitants.

Another activity that would encourage students to pay close attention to details and then draw conclusions would be to have them write a poem based on the details in the hut, the village, the son, the arrival, the meal—then students could draw inferences on each topic before combining their overall findings. What is the author's point? Is it quite clear? But the more pressing questions remain and deserve to be asked. In fact, students should be encouraged to raise three or four questions about this piece themselves. One hopes they would ask the following:

Why might the son have stayed away for forty long years?

What reasons might there be for the village to remain so simple, so much the same?

Where does one find happiness and contentment?

And is the son happy and content? Can we tell?

In what ways does tradition protect values? In what ways does it fail to recognize the importance of change?

What is the most necessary to a good life?

In the end, this piece should raise more questions than it answers. The theme seems to ask readers to draw a conclusion that contrasts sharply with the way many Americans live their lives. Ultimately, the story asks us, as readers, to consider how we live our lives and to identify the way we place value on our wants and needs.

FAMILIES

Photographs

1. When you look at the first photograph, what is your initial impression? What is the occasion for this gathering? What are the relationships between the individuals in the photos? How do you see meals—eating together—as related to family celebrations and connections?

2. What might be the relationship among the four people in this picture? Who are the two men? Brothers? Friends? Lovers? What details lead you to this conclusion?

Tran Thi Nga, Letter to My Mother (text p. 106)

Suggested Cross-Reference Readings	*Text Pages*
The Promised Land	62
The Son from America	95
Mute in an English-Only World	151

Although this piece is short, it is full of details that show the contrasts between two very different worlds. To encourage readers' appreciation of the author's experiences, the pre-reading questions ask students to think about an older family member who has had a positive influence on them, the sacrifices that people make for one another, and what students would do for their parents if they had the chance.

From the opening sentence, students will see that the author misses her mother when she writes, "You are always on my mind." The letter continues in a reportlike manner until paragraph 7. Here, the author stops relating the details of her life in the United States and goes back to her concern for her mother. The daughter is worried, anxious, and sorrowful without her mother, without her homeland. At the same time, she is grateful for all that she has— her own family, home, and life of opportunity (suggestion 1).

After students read this piece, ask them to recall, without looking back at the text, what details most impressed them. Chances are that they'll remember the ice cream, perhaps even the flavors. They may also remember colors and plants that Nga mentions. If this happens, the purpose of using details is clear: It makes the writing come alive. It gives the reader special pictures to take away. It connects the reader to the written experience (suggestion 2).

Although Nga details the vegetation, colors, and places she's seen, she does not mention any specific friends or neighbors. (She does mention "people from the church" who came and took them to "Friendly's for ice cream," but the people are faceless, and the visit seems like a duty instead of a friendly visit.) (suggestion 3)

Ask students why Nga seems so alone; remind them to think about the time that Nga and her family have been in this environment (three years, paragraph 3). What Nga doesn't say here, then, becomes a vital commentary on how isolated (ignored?) immigrants can be, even after several years of living in one place.

Most students will comment that Nga will most likely never see her mother again. After all, it's already been over three years, and Nga has not even heard from her mother. Students may question whether the mother is still alive.

The final metaphor Nga uses is powerful and poignant. The pond, whether small or dirty, is a place that is whole, solid, together. It may not be fancy, it may not move swiftly, but it is united and together. It is how Nga wishes her life to be: whole with her family, including her mother, in a safe, simple spot.

Students may enjoy writing from the mother's point of view. Although this seems like a simple exercise, if students try to put themselves in the shoes of someone who has never seen this country, the letters they write will motivate them to raise many questions about the culture in which they live and which they may take for granted.

Doris Kearns Goodwin, From Father, with Love (text p. 109)

This selection, an excerpt from Goodwin's best selling book, *Wait Till Next Year,* proves a favorite of most students, especially those who love baseball, America in the 1950s, and childhood memories that link them to an especially loved person. Having taught this piece several times, I've come to expect several students in each class to purchase Goodwin's book, anxious to hear more of her moving tale of growing up in Brooklyn.

Known for her simple, engaging style and her down-to-earth wit, Goodwin shows students how an ordinary topic—a hobby shared with a parent—can connect to a wide audience of readers simply by capturing them with details that slow down the action. In general, student writers have a tendency to move much too quickly when they write about a subject, so this piece offers a fine example of how the writer's camera can—and very often should—stop and shoot in one place. Paragraphs 4, 10, and 11 provide especially useful examples of this writer's technique.

While discussing this essay, it's easy to raise students' awareness of the purpose for slowing down and "showing" the action. While most students know they should "show instead of tell," many don't stop to think why. With Goodwin's story of her father, readers can see how the details of the piece make it absolutely believable, and how the writer thus becomes a credible, trusted ally.

Of course, some developing writers may not find this part of writing so intriguing. They will, however, be able to relate to the topic at hand: the memories of one's early years and the way life used to be when one was younger. Students might respond well to the following prompts to get them thinking about their own connections with the essay: What one summer in your life stands out as the most memorable? If you picture yourself and a parent when you were young, what one picture or scene comes to mind? If you could hold onto any one aspect of your childhood, what would it be—and why?

All of these questions, of course, should then be asked about the writer: Why is this one summer so memorable? What one picture comes to mind of her and her father? What one thing would she like to hold onto? Can she? Can we?

Finally, while this essay is certainly entertaining on a nostalgic level, it serves also as a model of a well-structured personal essay. For example, Goodwin indicates her thesis with the opening line of this selection. It's carefully crafted to reveal three separate, yet connected, components, which she will expand in the body of the reminiscence. Students should easily see how clear Goodwin's plan is.

In addition, they might sketch out similar plans of their own, perhaps as an in-class exercise. They could, in fact, use Goodwin's structure as a model, replacing it with their own truths: "The _____ has always been linked in my _____ with _____, and _____. The results of this exercise could simply be discussed in class or in small groups, or students' statements could serve as the beginning of a longer essay for an out-of-class writing assignment.

Gary Soto, Like Mexicans (text p. 113)

Suggested Cross-Reference Readings	*Text Pages*
Spanglish	138
Finishing School	179
A Question of Language	142

When the question is posed, most students will insist that they are not prejudiced, firmly stating that they treat everyone the same. Before or even after reading this piece, it might be a good in-class exercise to have students list all of their close friends whom they would not consider to be in the same culture group as they are. In doing this exercise, I assure students that I will not collect their lists, nor will I insist that they share what they have written. Nonetheless, as students consider their own lists, what will probably emerge is this: although most of us claim that we're not prejudiced, few of us have a great number of close friends outside of our own culture groups.

As the exercise shows, Gary Soto's experience was not unique. In fact, it was and is quite common. Students should note all the different people who shape Soto's views: First, of course, his family members and then his close friend. From these two perspectives, he draws the conclusion that he, too, will marry a Mexican girl, someone who is of his own culture.

Students will also find plenty of evidence in this piece that good writing contains good details. The colors are vivid, as are the details of the homes. In fact, the environment garners more description than the people in this piece. An intriguing point of discussion might be to consider why Soto focuses so strongly on the things that surround him. (His point could very well be that it is our early environments that shape who we are to become, and indeed, if it weren't for Soto's educational pursuits that took him away from home, he might never have changed either.)

Students, too, could take note of the way this piece is organized. It goes in chronological order, with an introduction that hints what the conflict might be. (Curious students may want to discuss what that "third" point is the grandmother never mentions at the end of the introduction.) The last line is one that deserves attention as well. What does the author mean that Carolyn was like Mexican but different? What is his point here?

One in-class exercise that works well especially at the beginning of the year when students do not yet know one another is to put students in groups of three and ask them to come up with three somewhat unusual experiences, traits, or similarities that they all share. (In other words, being students or being female or having two eyes doesn't count here!) Initially, students don't think they can find anything at all that they share. However, after about ten minutes of probing, students find more in common than they ever thought possible. This exercise reinforces what Soto discovered in his own life—that as unlike each other as people may seem, if we take the time to know each other, we find that our similarities far exceed our differences.

Lindsy Van Gelder, Marriage as a Restricted Club (text p. 118)

Suggested Cross-Reference Readings	*Text Pages*
Parish Streets	85
The Loudest Voice	211
Unfair Game	282

The first three selections in this section center around personal stories that are based on childhood experiences. While all the "family" selections deal with struggles in one way or another, this is the first piece that deals with an issue some students may find controversial: gay marriages. Some students may not see why this selection is included in Section 7; however, after some discussion, most readers will agree that it fits into this category under the following definition of family: the people with whom we live or are connected by special bonds of love and physical closeness.

Whether or not students agree with the author's premise here, they will find this piece clear and direct. Unlike other selections in this section, this piece `is arranged primarily as an argument—not as an informative or persuasive piece of writing. Students will find the thesis or point in its traditional place: at the end of the first paragraph. From there, the author builds her argument for her position. Students should examine the reasons the author presents. How reasonable do they find these arguments? Do they find any that are unreasonable? How fair do they find the author's "voice"? In other words, how does she present herself, and how might she view her audience?

Students could also respond to the places in this piece that interest them the most and the places where they start drifting. (Many students complain that once they get to facts, such as in paragraphs 4, 5, and 9, they get "bored." However, it is important to stress that a good argument, indeed, must be grounded in facts rather than only in feelings or opinions if it is to be strong.)

Certainly the question that Van Gelder indirectly poses should spur additional discussion: What issues and questions are you willing to consider seriously, whether or not they directly affect you? While many students may see this as only a gay or lesbian issue, at its root, it is not. It is a question of moral responsibility and the courage to take a stand when it is not popular. Another question bears asking as well: Is this piece relevant today, seeing that it was originally written over fifteen years ago? Do students think views on gay or lesbian marriages have changed since 1984? What proof (facts) could they get to help them answer this last question?

Digging up answers for this last question might serve as a good introduction to research strategies and the use of sources as support for one's point.

Sue Horton, Mothers, Sons, and the Gangs (text p. 123)

Suggested Cross-Reference Readings	*Text Pages*
The Promised Land	62
I Show a Child What Is Possible	170
Prison Studies	206
How Boys Become Men	268

Moving students toward responses that require a bit of distance as well as support, the pre-reading questions ask students to describe, explain, and basically freewrite an argument. Questions 2 and 3 would work especially well in small group brainstorming sessions. Groups could then present their findings to the whole class, which could summarize the findings. Thus students work through a whole thinking process of reacting, explaining, and summarizing.

To reinforce how vital good support is to any argument (thesis), suggestion 1 asks students to search the text, respond, and then explain their responses. Certainly personal experience will come into play here, but if you have the time, push students beyond their own experiences to other sources such as current events, a current movie, TV show, article, or personal observation.

According to dictionary definitions, *machismo,* which has the same roots as the word *masculine,* means an exaggerated masculinity, emphasizing physical courage, domination, and aggressiveness. Students could also look up the following words to get an even deeper meaning: *masculine, manly, macho, manhood.* Clearly, students should see that all of these words focus on an exaggerated effect of power and control (suggestion 2).

Again, if students can remember one character or one scene, then the writer has done her job. If students are sharing writing at this point in the semester, you may also want them to point out the memorable parts in each other's drafts (suggestion 4).

Question 4 fits snugly with the second pre-reading question. Expect a great deal of discussion on this question. At first ask students to volunteer personal experience and observations. Then transfer their attention to any experts they might know in the field.

As an alternative, consider putting one character from this piece on trial, dividing the class into jury, judge, defenders, and prosecutors. Give students adequate time to develop their major points, and then let the jury decide the verdict, with the judge pronouncing the sentence and giving the reasons for it (suggestion 4).

Students should be able to gather, from the title alone, that few of the gang members in this piece have fathers or other positive male role models. You may ask students to explain this phenomenon. They might also gather statistics on the number of female heads-of-household in this country.

To pursue this issue, you could ask students if they believe that boys without fathers are more likely to join gangs. (The Rodriguez boy has a father, who is mentioned only once. Have students find that section and ask them to draw inferences about the influence the father has or doesn't have on the family. What difference does it make that he comes home late at night from work?) What other factors do these stories have in common, besides the absence of a father?

The concept of "in" groups and "out" groups is very real, not only to this piece but to the cultures most of us live in. Having students first make a visual map of the cultures in which they fit—and the cultures from which they are automatically excluded—might be helpful before delving into gang cultures. Consider having members (or ex-members) of a local gang talk to the class about their experiences. Also consider having parents of gang members come and speak to your class. Some class members may have had experiences with gang culture that they are willing to share. Showing a movie on gang experiences, such as *Boyz 'N the Hood,* might be a powerful classroom experience as well.

Having students draw up charts with headings such as "Parents," "Siblings," "Education," "Neighborhood," "Personality," and "Drug and Alcohol Use" and then fill in examples for the six categories will help students see the differences, as well as the similarities, among the three gang members. Based on this visual comparison, students should be able to draw inferences about what prompts different individuals to join gangs (suggestion 7).

Now that students have thought about the gang culture as described by Horton, they will be able to write about their own teenage culture. The visuals they have made may help them.

Besides explaining their cultural experience, students could also write about one time when their parents gave them advice, or they could write a letter to one of the gang member's parents explaining their own point of view and perhaps giving advice.

Seamus Heaney, Midterm Break (text p. 134)

Suggested Cross-Reference Readings	*Text Pages*
A Brother's Dreams	221
How Boys Become Men	268
Nurses in Vietnam	242

To interest students in this powerful poem, you might begin by asking that students read the poem straight through on their own. Then they might freewrite for five minutes or so on anything that touches them about this poem. What words do they like? What is confusing? What makes sense? From a first read, what would they say this poem is about? Most students will see that this poem is about a student who leaves school in order to come home to his young brother's funeral.

Now, you might read the poem aloud slowly and carefully. After this reading, students might jot down any new responses or insights. A discussion on these new insights often reveals the value of reading and rereading as well as the power that comes from hearing poetry spoken aloud.

The next step might be to go back again—to look at specific images. Which ones do the students find most memorable? Which ones touch them the most? In addition, the questions should start popping up. Who are all the characters in this piece, and why do they matter? How do their reactions to this death compare to one another's?

Again, they should be asked to reflect: With these images in mind now, what might be the author's message about this topic of death? Students will not be able to get to this, however, until they look one more time at the speaker. Who is he? What does he notice? How does he react? Just as in a short story, the conflict in this poem centers around this young student.

If students examine, paragraph by paragraph, the reactions and insights of the speaker, they may glimpse a connection with this poem that goes beyond the sorrow connected with death and loss. Through the speaker, we see the process that we, as humans, tend to go through: The initial sense of denial, concentrating on the daily routines. The reactions of others, the self-consciousness of being consoled, the place where strangers come in, the centrality of the corpse, the way time seems measured and slow.

Finally, (stanza 6) in the morning, a time of day that suggests innocence and beginnings, the speaker goes into the room where his dead brother is laid out. In this stanza, the images are soft, pure, and elemental, relating, for example, to flames and water. The speaker finally looks. He finally sees. In fact, he sees "him"—the first mention of this corpse as a person. The speaker is close now, and sees the "poppy" bruise—like a flower. Still being exact, he notices it is on the left temple, "his" left temple. The last stanza, unlike all the others, is only one line, providing emphasis to the poignance of the speaker's loss.

QUESTIONS OF LANGUAGE

Photographs

1. In the first photograph, can you tell which woman is learning a language and which one is teaching it? Do the women seem to understand each other? What details lead you to your conclusion?

2. In the second photograph, what do you see as the significance of the Statue of Liberty? Why would these protestors choose the statue as a symbol to promote their cause?

3. In the second photograph, one of the banners proclaims that "Censorship Is Un-American." What does this slogan mean to you? How would you define censorship? Do you see it as un-American? Explain.

Janice Castro with Dan Cook and Cristina Garcia, Spanglish (text p. 138)

Suggested Cross-Reference Readings	Text Pages
Like Mexicans	113
The Son from America	95
Parish Streets	85

This short piece provides a fine way to encourage students' listening skills. Students may have a tendency to skim over this piece, getting the context rather easily, without making the connections the authors desire. You might point out, however, that language is a changing, vibrant part of our lives, and if they listen closely to the way they and others speak, it's usually easy to see how many words in our language are shortened, combined, and reduced to slang in our attempts to communicate with one another.

As these authors also point out, right from the start, language can bridge generations as well as connect strangers in crowded cities. How does our language change to meet the needs of our changing world? Again, as this article suggests, the change in population as well as the change in media are two important forces that have led to the increase of Spanish root words and phrases in the American vocabulary.

To listen to the vibrancy of language in our everyday lives, students could have short group projects in which they do the following: Listen for "Spanglish" phrases both in their personal encounters as well as in the media—radio, television, newspapers. Students could also form groups that agree to research the current influence of the Spanish language and culture on the current foods, music, and recreational areas in America.

Of course, students could also listen for influences of other languages that have entered the mainstream of English. This listening exercise can last over the course of a few days or, alternatively, as a semester-long extended writing and discovery project.

Gloria Naylor, A Question of Language (text p. 142)

Naylor uses her personal experience as a way to draw inferences about how much spoken language affects the way we feel and think. Students are asked to consider their own experiences with words before reading this piece. Some students may be open to sharing their list of "forbidden" words, and if so, such an informal discussion will further connect the students to Naylor's main points as well as to her process of presenting her arguments.

After reading this piece, most students will recognize that the controversy surrounding language is not so much what is said but how and why something is said. Naylor suggests that many words that are regarded as offensive are not in and of themselves "bad" or "good." They are only tools that are used to serve a speaker's purpose. And when that purpose is to divide or hurt instead of heal and unite, language can be bitterly destructive (suggestion 1). Affirming this point, Naylor notes, "Words themselves are innocuous." Thus the power to move people comes not in the words but in the general agreement about how the words are used and what is meant by them (suggestion 2).

Naylor then uses one particular word, nigger, to demonstrate her thesis. She describes situations when black people speak among themselves and use this word; it can then be innocuous, as long as the group accepts the meaning in a lighthearted way. It may even be a term that suggests unity among black people. However, when the third-grade white boy called Naylor a "nigger," the term became destructive for it carried the implication that she was something dirty, less than human (suggestion 3).

Students may be able to reconcile this apparent contradiction if they look to how their own families or different groups they are in use language. For example, members of a family often share jokes and nicknames. However, if someone outside the family uses the same teasing term, it takes on a new meaning (suggestion 5).

What does a parent say to a child who has been verbally attacked by another child? Consider asking two volunteers to role play a possible conversation between the young Naylor and her mother or father as they discuss what has happened at school that day. Students observing the role playing should pay special attention to the words that the two speakers emphasize (suggestion 6).

As an extended activity, students can conduct a survey to discover the subtle connotations of certain "loaded" words. Suggest that students have respondents use words in sentences as well as give definitions so that the true connotations can be unraveled. Surveying and collecting material might be a productive small group or even whole class activity.

Barbara Ehrenreich, Zipped Lips (text p. 147)

Suggested Cross-Reference Readings	*Text Pages*
Marriage as a Restricted Club	118
The Son from America	95
Finishing School	179
The Use of Force	254

This selection, like "Marriage as a Restricted Club" by Lindsy Van Gelder, develops an argument. As she builds her case, gifted essayist Ehrenreich keeps her audience's attention by developing this argument through the use of vivid examples and anecdotes.

Students can learn a great deal about their own beliefs from reading this piece, as well as a great deal about what constitutes good writing. On a personal level, this essay raises the question, How free are you in America? Students should offer their own experiences in the workplace to support their responses. Many will identify with Ehrenreich's thesis: that an employee is, in many ways, under the restriction of his or her employer. Do students think this is unconstitutional? Perhaps many have never thought of the question in this way before. Again, discussion should be lively around this type of topic, which virtually all students will find familiar.

Second, students should be made aware of all the elements of good writing that Ehrenreich uses in this essay. For starters, what about her introduction? How effective do students find it? Why? One might want to point out how details add life and credibility to a piece of writing. For example, Ehrenreich doesn't just say, "some guy" in her introduction, but she gives him a name: "Sam Young." By doing so, the character becomes concrete and human. The audience cares more now.

From this opening example, the author continues to support her argument with specific examples and details. These examples serve to pull the audience in before she gets down to the nitty-gritty of her argument (paragraphs 7 and 8). And although this is a formal argument, students should be aware of the "voice" that this author chooses to adopt. As opposed to a strictly formal tone, she writes in a colloquial manner, interspersing slang with more traditional word choices.

If one of the course goals is to encourage students to read slowly and closely, then this piece offers a great opportunity for practice. (Robert Frost made a good point when he said that perhaps the main reason students need college is to relearn how to read.) Ehrenreich's essay works on many levels. Students can read it for pleasure and for writing responses. They can also analyze the writer's tools Ehrenreich so skillfully employs here. In addition, they can read it for content and meaning, pausing to understand all the vocabulary, and working on writing a paragraph-by-paragraph summary that shows they have grasped the essence of this piece and can retell it effectively to others in an objective way.

Finally, students can write their own pieces, either arguments about their own rights, or narratives that tell the story of when their rights or someone else's were infringed upon. Another writing option might be to define powerful and complex words such as *freedom, democracy, dictatorship,* and *citizen.*

Chang-Rae Lee, Mute in an English-Only World (text p. 151)

Suggested Cross-Reference Readings	*Text Pages*
Letter to My Mother	106
Fitting In	81
Modern Courtesy	310
Like Mexicans	113

This is a powerful piece on how language can shame and humiliate the person who is not skilled in the major language of a culture. It is, in many parts, highly vivid and detailed, especially in the encounter the author relives when he was a very young boy.

In order to encourage students to connect with this piece, it would be valuable to discuss the concepts of "insiders" and "outsiders." In what groups do they feel most comfortable? In what groups would they feel most awkward and self-conscious? While many students may not have directly experienced language as a barrier, they certainly will be able to identify other social barriers that have excluded them from belonging in one sense or another.

While the theme of being an outsider is clearly evident in this piece, there are other issues worth discussing. First, there is the role of the mother in this piece and the type of person she is. Most students will surely admire her for her grace and courage. This writing may also elicit anger when the mother comes in contact with "one of the beefy men" behind the counter. Students will be justified in their anger at this point. This is the perfect opportunity to discuss how we treat others when we're in a hurry. How do we act if a customer of ours is slowly counting out change? How do we behave if the driver of a car in front of us seems lost and is driving slowly? Although it is our first reaction to chastise the "beefy" man, to be fair, we need to look at our own reactions as well. Perhaps the question deserves to be raised, What are the reasons so many people react from time to time with impatience and rudeness?

Another point worth discussing or writing about is the issue of how people learn. What are the steps one must go through in order to learn? What factors tend to motivate people to learn? What roadblocks can one anticipate? Students should also feel free to share their own learning experiences—their successes as well as their failures. As an additional assignment, they may want to write an essay on how they got so good at one particular skill, detailing the ingredients for success as well as the pitfalls they encountered.

This essay is carefully written and exacts from the reader a variety of reactions, not the least of which, one hopes, is compassion. If you would like to reinforce this quality, then the extended writing suggestion—to interview someone who has been a stranger in a country other than his or her own—would be a fine way to deepen the connection and meaning of Lee's theme.

Joseph Telushkin, Words That Hurt, Words That Heal: How to Choose Your Words Wisely and Well (text p. 155)

Suggested Cross-Reference Readings	*Text Pages*
Finishing School	179
Prison Studies	206
Unfair Game	282
Modern Courtesy	310

Although this piece is long, most students easily grasp Teleushkin's ideas. The essay, which is primarily informative, divides into categories so that discussions and writing can be linked to each category and to the subthemes of these sections. The beginning exercise in this essay—that of asking people if they can go twenty-four hours without saying an unkind word about anyone—provides a good starting point for class. Since Telushkin predicts most people cannot go this long, he makes the connection that people are addicted to harmful language. What do the students think about this point? Is his contention just?

The next section in this piece provides a thoughtful out-of-class writing activity, one that should raise the writer's awareness of his or her own habits.

Most students will actually enjoy doing this type of objective research, finding out, as it were, how often they say something negative about someone else. These findings certainly should be shared in class after the observations are made.

Having made his point that most people do indeed use language as a way to harm another person, Telushkin then proceeds to talk about the power of language—a power to which many remain oblivious. In this section, students will be able to recognize the authorities to whom the author appeals for support. If the instructor wishes, this would be a fine time to introduce the three types of appeals that those who want to persuade or inform have at their fingertips: logical appeals, emotional appeals, the moral or ethical appeals.

From here, the students are faced with language as an ethical issue. Students might need to slow this section down, taking the time to understand the vocabulary as well as the main point in each paragraph. Students could paraphrase the meanings within each paragraph and then compare each other's summaries to see if they are getting the main ideas here. Once the author's message is clear—that even passive listeners have a moral obligation to speak up when derogatory comments are made—then the argument can ensue: Do students agree that their silence is unethical in such cases? How strong is the author's argument here? Can students find a counterargument to his position?

Telushkin continues with other points about language: the evil of slander and the responsibility one often has not to spread news, even if it is the truth. He ends with a proposal: a national "Speak No Evil" Day. What do students think about this idea? Would they be willing to sponsor such a day on their own campus? Why or why not?

After reading this piece, students may want to reflect on the content of Telushkin's message, and non-Jewish students might wish to recap what they have learned about the Jewish religion that they didn't know before. They could compare the tenets listed here with tenets of their own or other religions. Such a research project should indeed be encouraged if the student shows an interest in such a worthy topic.

Emily Dickinson, Three Poems on Words (text p. 164)

Suggested Cross-Reference Readings	*Text Pages*
The Son from America	95
Midterm Break	134
"Real" Men and Women	279
Harrison Bergeron	327

In no other genre is the word more important than in poetry. In words and through the arrangement of those words on the page, poetry presents its meaning to the reader. Of course, not all students using this text will aspire to be Dickinson scholars; however, these three short poems introduce readers to the simplicity and complexity that lives in this classic American poet's work.

As in the suggestions for reading poetry given for "Midterm Break" (text, p. 134), readers will benefit from focusing on specific words and their denotations as well as their connotations. Students may or may not know the difference between these two terms, so it might be a good idea to take some concrete images in each poem and approach each first definitively and then analytically.

Groups of students could each work on a poem and its images and present their findings and connections to the class. They should approach this exercise as if they are uncovering every possible meaning and connection that the words themselves hold. In addition to the words, they may even think about the lines and the rhythm.

It may be a good idea, too, to tell students in advance that many people find Dickinson's poetry elusive, while others find each poem to be like a flower: One gazes on it time and again, yet it is often hard to describe or to explain the feeling one gets from it. Students should try, though, to at least expand on the images, for they are, though simple, also full of depth and beauty. The following is a partial list of possible words students could examine in their groups:

melody
silent
communion Wine
native
Blades
Nerve
Bone
Flesh
Creatures

In addition to exploring these three poems and what they have to say about language and the power of words, students may opt, as a creative writing assignment, to go back to their own pre-reading lists, in which they wrote twenty of their favorite words, and work on their own poems, perhaps using Dickinson's simple and stark style as a model. Any students who are moved or intrigued by the three poems in this text should be encouraged to read more of Dickinson's poetry and perhaps even center a semester research project around this famous and controversial American author.

WAYS OF LEARNING

Photographs

1. In the first photograph, what does the act of painting seem to mean to the woman? What are three possible reasons why this woman appears so pleased? As you develop your explanation, create an image of her painting (in words or through painting or sketching).

2. In the second photograph, who is the man who is speaking? A teacher the students see every day? An invited guest? What is he wearing? What impact do you believe his appearance (or the appearance of any teacher) has on the learning experience of the students?

Jacques d'Amboise, I Show a Child What Is Possible (text p. 170)

Suggested Cross-Reference Readings   *Text Pages*

The Woman Who Makes Swell Doughnuts	68
Mothers, Sons, and the Gangs	123
How Boys Become Men	268

Continuing the theme of how we can achieve our greatest potential if we are fortunate enough to have a mentor along the way, students can write or discuss how they have learned to do something well. (Students should know that they can be experts at almost anything. One student proudly said she was an expert on eavesdropping and conducted a miniworkshop to teach other students her "skill!")

Although few, if any, students in your class will be able to identify with the grand success that d'Amboise enjoyed as a dancer, there are certainly spots where almost all of us can connect, especially when the author writes about the importance of praise in motivating students. Most students will be able to think of specific examples when a teacher's comments were the turning point—to either success or failure.

Many students will also be able to identify with how paralyzing fear can be when we're learning something new. Being able to take the first new step, figure out the word problem, play a musical note clearly, do a front flip off a diving board—any experience in which we are able to do something new—sends us soaring as students (suggestion 1).

To add another dimension to this experience, you might ask students to think about the last time they had this "soaring" feeling. Then have them draw conclusions based on their responses. Does this feeling die when we start formal education? Do we become too old to need this feeling?

Like other teachers depicted in this section, Madame Seda knew how to channel a child's energy. She also knew how to capture students' interest and make them want to learn. In addition, she knew the importance of discipline and setting limits and consequences in her classroom. If students have read other pieces from this section, have them get in groups and compare Madame Seda to Miss Hurd, Jack MacFarland, and the grandmother in Casa, as well as their own teachers (suggestions 3 and 4).

As students continue analyzing what makes a good teacher, they should think about the types of teachers they would make! How many of them would go into education? Why? How many wouldn't? Why not? Besides external factors like salaries and working environment, what personal qualities would keep them in, or out, of the classroom (suggestion 5)? Students might also consider ways in which they will be teachers whether or not they formally pursue this profession.

In this piece, d'Amboise feels a teacher should be three things: precise, clear, and truthful. Students may also see that teachers need to be knowledgeable, caring, committed, capable, and fair. They may think of other desirable qualities as well. If you want a debate to ensue, ask students what single quality is the most important and have them support their response with specific examples (suggestion 6).

Since the student is a central factor in whether a teacher is successful, students could write an extended essay on a particular process they have gone through to learn something new. Organizing a process essay into a series of chronological steps is fairly easy, yet presenting exact steps and details challenges many students.

D'Amboise's image of each person as a "trunk that's up in the attic" invites students to explore this metaphor as deeply as they can. The two key components here, of course, are the *trunk* and the *attic*. What do trunk and attic imply? (A trunk is a place to keep important things, a heavy piece of furniture that's hard to open; it's often locked; it's often "handed down"; its essence is inside; it can look quite ordinary on the outside. The attic metaphor can be extended this deeply as well.)

After extending this one metaphor, students may wish to come up with their own metaphors. Students may find it useful to describe an idea as a thing in a particular place.

Isaac Asimov, What Is Intelligence, Anyway? (text p. 176)

Suggested Cross-Reference Readings	Text Pages
Mute in an English-Only World	151
A Brother's Dreams	221
Of Headless Mice . . . and Men	301

Most readers find this selection to be a perfectly delightful piece. It is clear and engaging, and perhaps most important, it addresses an issue that is seldom raised by those deemed highly intelligent. What is intelligence, anyway? Is it simply a score on an aptitude test? Is it indicated by earning the highest grades in a class? Is it the student who can figure out an advanced calculus problem? Or is it, as Asimov so humbly and astutely suggests, any ability that comes to the rescue when creative resources and solutions are needed?

In this broad sense, if the toilet is clogged, a plumber's intelligence will be worth more than an astronomer's. Likewise, a dancer will prove more intelligent than a scientist if the problem is figuring out how to work three pirouettes into a dance routine.

Given this definition, most students will be willing to discuss this piece in terms of their own intelligence. The pre-reading questions set them up to think of intelligence as a one-dimensional, testable trait. Another question before reading asks them to weigh the value of knowing versus doing. It might be worthwhile to assign both of these pre-reading activities, perhaps one outside of class and one in class. Either way, these questions will spur discussion that moves in the direction of Asimov's thesis.

Since the piece is straightforward, you might consider assigning a fairly complex writing topic that builds on the reading, thinking, and discussion of this selection. Students have several fine choices for making writing connections after they have thought about this topic. They can brainstorm on the "smartest" person they know—and now, this brainstorming can take on new dimensions. In fact, as a whole-group activity, the class could brainstorm for categories for such a topic. Some of the categories might include the following:

People who . . .   work well with their hands

                        have common sense

                        are good with animals

                        are physically strong and able

                        can work with wood

                        can grow their own food

                        see the good in everyone

As you can see, the list is practically endless. However, it leads right into the second choice for extended writing: to research Gardner's theory on intelligences and report on his findings. In fact, if your students are ready for synthesis, you may combine these two choices into one. Looking at Gardner's work provides an extended project for all those who may be drawn into the question What is intelligence, anyway?

Maya Angelou, Finishing School (text p. 179)

Suggested Cross-Reference Readings	*Text Pages*
A Question of Language	142
Mute in an English-Only World	151
Three Poems on Words	164
Just Walk on By: A Black Man Ponders His Power to Alter Public Space	262

In general, students thoroughly enjoy this story and identify strongly with Margaret. You might want to remind students that Maya Angelou's birth name was Margaret (or Marguerite) Johnson. Students usually enjoy discussing and writing about their own names and explaining what names they like to be called, what names they tolerate, and what names make their blood boil. If a discussion on the importance of names happens
before students read this piece, they will more fully understand Margaret's response when Mrs. Cullinan insists on calling her Mary.

Selecting a significant educational experience outside the classroom to write about is usually quite easy as well. This pre-reading topic also makes a nice extended writing assignment.

Most students will say they like this story. If you ask why, many note the action, the fast pace, and the dramatic climax when Margaret breaks the china. Others might move to another level and talk about Angelou's style: the fine description, the natural dialogue, the honest voice of the narrator herself.

On a literal level, Mrs. Cullinan is "barren," meaning she cannot have children. On a symbolic level, her infertility also means that nothing new and promising grows within Mrs. Cullinan mentally, socially, or spiritually (suggestion 2).

Certainly the idea that everything and everyone has its place is the rule in Mrs. Cullinan's household. Whereas she sees this as order and elegance, others see her often cruel distinctions as discrimination. Students may be able to see parallel connections between where people sit at formal affairs, the behavior of members at exclusive country clubs, and the placement of desks in a classroom. Are these situations examples of order or discrimination? Students should see that this issue is not as simple as we might see it in this story (suggestion 3).

Young Margaret seems like a warm, loving person. In the beginning, she feels sorry for Mrs. Cullinan in her barren state. However, when Mrs. Cullinan takes it upon herself to change Margaret's name, the child no longer feels sorry for this white woman. As a matter of fact, she is so angry she thinks about writing a poem to mock her (suggestion 4). This shows us that Margaret is willing to give anyone a chance but will not allow herself to be abused. She shows herself to be compassionate yet strong, with a clear sense of self-worth.

In a way, Miss Glory and Margaret are the choices we all have when we are faced with injustice. We can either cope and stay quiet (Miss Glory), or we can think and react to the situation. Although most of us admire Margaret more than Miss Glory, we might have to admit that when faced with opposing authorities who hold power over our lives, most of us do what's good for us rather than making waves (suggestion 5).

The "Alice in Wonderland" reference in this piece works on several levels. The world into which Alice falls when she stumbles into the rabbit hole makes no sense. Those who hold power hold it for no good reason, and they abuse their power in astonishing ways. Mrs. Cullinan is cast in the role of the ugly, spiteful Queen of Hearts (further irony) who controls everything and everyone around her. Margaret is seen as Alice, the one who questions, does her best, and eventually rebels, revealing the absurdity and unfairness of the Queen's world. Further play with this analogy connects Glory to the nervous White Hare, Bailey becomes the wise caterpillar, and the dish-breaking ceremony parallels the Mad Hatter's tea party (suggestion 5)!

Mrs. Cullinan shortens Margaret's name as a means of control. In a sense her action means Margaret belongs to her household (suggestion 7). Bailey, the wise and inconspicuous cohort, offers Margaret a way of getting even with Mrs. Cullinan. Since Mrs. Cullinan destroyed Margaret's heritage, Margaret will destroy pieces of hers.

Could Margaret have solved this dilemma any other way? What if she spoke up to Mrs. Cullinan? Couldn't she have said she would not answer to the name Mary? Couldn't she have said she would quit unless her employer called her by her right name? Students might ponder whether these choices might not have been more acceptable. Many students will see, though, that since Mrs. Cullinan is "inhuman"—or at least oblivious—appealing to her rational, sensible human side will not work. In order to smash the control Mrs. Cullinan has, Margaret has to smash something her torturer values.

The extended topics offer possibilities both for personal writing and for objective observation. Writing about who we are is not always such an easy task. However, if students choose an event and write about how they reacted, they may be able to see part of who they are in a clearer, more objective light.

As another writing choice, students can take the spotlight off themselves and observe someone else in action. Dialogue as well as body language are important cues here. If students have vivid memories of a mean-spirited person in their own lives (and somehow, the villains always remain vividly with us), they might enjoy writing a description of this person. After all, as Margaret discovered, revenge can provide great motivation for writing!

Nicholas Gage, The Teacher Who Changed My Life (text p. 186)

Suggested Cross-Reference Readings	*Text Pages*
The Promised Land	62
The Education of Berenice Belizaire	72
From Father, with Love	109

If students are also reading the Rose piece (text p. 193) this essay continues the discussion on the influence a teacher can have on a student. If they haven't already discussed an important mentor in their lives or the characteristics they believe a good teacher should possess, then writing on these topics in their journals would be quite valuable.

Students can also write about their own experiences with writing. This reflection works well at the beginning of the semester, but it is equally valuable at any point because it helps students see the relationship between process and successes in writing.

Besides linking up with a wonderful teacher, Gage is successful in his writing for other reasons. First, he takes the advice he is given seriously. He listens, learns, and tries. He also is able to write on a topic that is dear to him. In this process, he doesn't worry about the grade he will receive; he doesn't worry about impressing anyone. He simply lets the writing take over, and, as a result, he writes honestly—something that is hard for most of us to do, especially on a first draft (suggestion 1).

Writing about his mother is a great outlet for Gage. As he says, he wasn't close to his father, who was quite strict, the ultimate authority in the house. Thus the assignment he was given was also a blessing for him, as he was able to get in touch with the emotions he was feeling and the past he needed to understand (suggestion 2). Most students would agree that they are apt to do better when they write about their own experience (suggestion 4).

Certainly Miss Hurd was wise enough to set up an assignment that would lead to success, if the student thought and wrote hard enough. Students should think back to teachers they've had who also possessed the ability to encourage the best in their students (suggestions 3 and 6).

Reflecting on how a particular season affects them is another writing topic that inspires most students. Like Gage, a student wrote how the fall reminded him of his mother, who died when he was young. Another student associated winter and the Christmas season with memories of her grandmother. Although writing of this sort can conjure up painful memories, if students can write through the pain, they are usually grateful for the experience (suggestion 5).

The teacher as catalyst is an apt metaphor in many ways. A catalyst is the agent that puts energy into motion and gets a process moving. If students glance back at their pre-reading responses to the characteristics of a good teacher, they should be able to create their own metaphors, based on their chosen characteristics. Encourage them to take one more step and ask them to come up with a metaphor for an excellent student, a good student, and a poor student (suggestion 7).

Having students send letters to editors of local newspapers is an assignment that usually has great success. For one thing, students have a real purpose for writing. In addition, many of them will see their names in print. For students who want to try more challenging publishing opportunities, suggest they write to a popular magazine.

As another real-world writing opportunity, students could write a letter that publicly or privately thanks a former teacher.

Mike Rose, "I Just Wanna Be Average" (text p. 193)

Suggested Cross-Reference Readings	*Text Pages*
The Education of Berenice Belizaire	72
Mothers, Sons, and the Gangs	123
Surfing's Up and Grades Are Down	305

It's guaranteed that nearly every student will identify with Rose's piece for one central reason: labels. Mention this word in conjunction with school, and practically all students will be able to tell you about students they knew (often themselves) who were adversely affected because of labels they were given in school.

With this focus in mind, the three pre-reading questions get students prepared to think about any labels they might have been given and about stereotypes they themselves had applied to others. They have the opportunity to think about their earlier school experiences and to consider what, if anything, they would change about their past educational lives.

Few students find this piece boring. Rose's lively descriptions and finely tuned bits of dialogue bring to life the teachers and students who inhabited his childhood and teenage classrooms. In fact, most students get so caught up in this piece that they start substituting people they knew for the individuals Rose describes (suggestion 1).

Many students will also have an opinion on standardized tests; if you ask them to describe their own experiences with such tests, expect many of them to say their experiences were negative (suggestion 2). At this point, you may want to ask students whether they think such a test is a good measure of their abilities and potential. They'll certainly have strong opinions on this one (suggestion 2).

Students also have strong feelings on levels and groupings in education. Some students may not understand exactly what "homogeneous" grouping or "tracking" means. Take some time to explain this concept and have students use their own experiences as support for their responses (suggestion 3).

If we were to generalize about the teachers in Rose's story, we could attribute the following characteristics to the teachers he mentions: Brother Dill—paranoid, insecure, and immature; Mr. Mitropetros—inadequate, a time-waster; Mr. Montez—incapable of controlling a class; Brother Clint—good, caring, an enabler of students; Jack MacFarland—master teacher, knowledgeable, enthusiastic, a motivator, a mentor.

Students may agree that they've had teachers with these same characteristics. Give them a push and ask them if Rose's ratio holds true: For every five teachers they remember, three are bad and two good (suggestion 4)?

It should be easy for students to come up with the "rules of the game" in education. What do students in high school generally have to do in order to be successful? (They have to do the work!) What have students done to learn in "boring" courses? How does real learning differ from following rules to get the grade or pass the course (suggestion 5)? A discussion of the difference between intrinsic goals for learning and extrinsic goals often provides intriguing responses. Many students have never considered setting or identifying intrinsic goals and have spent their school lives judging their success solely by the evaluations of their teachers.

Finally, as students read about a master teacher, Jack MacFarland, they should also think back to their own experiences, to their own teachers. Which teachers inspired them? What courses were valuable to them? What methods of teaching do they think are best for students (suggestion 6)?

Now that students have reflected on their education up to this point, they should be set to do either of the extended writing topics: Many students are pleased to write back to their school or to a particular teacher, but some may prefer to look at how they have fared as learners. By comparing their experiences to Rose's, they may see their education in a whole new light.

Malcolm X, Prison Studies (text p. 206)

Suggested Cross-Reference Readings   *Text Pages*

Suggested Cross-Reference Readings	Text Pages
Crossing the Border	76
Mute in an English-Only World	151
Mothers, Sons, and the Gangs	123
Surfing's Up and Grades Are Down	305

As Robert Frost suggested, the primary reason to go to college is to learn to read. And for most students, whether developmental writes or seasoned ones, this philosophy seems to hold true. Most students need to learn to slow down their reading, to take time to digest what is being said to them. Unfortunately, though, many students set as their goal simply to skim as quickly as possible and to get a reading "over with" instead of considering the author's ideas thoughtfully and carefully.

You can discover your own students' reading process by asking them to do a quick writing about how they went about reading this piece. Were they in a room alone? Did they read it once and put it away? Or did they go back to it, raise questions, and try to discover the meaning of words they did not know from contextual clues or by looking them up in a dictionary? Did they then read the words they didn't know aloud? Did they try to use them in their own sentences? Did they discuss this piece with friends or talk to anyone about the issues "Prison Studies" raises?

Most probably, the writings will reveal that students read this piece quickly, while others were around. Perhaps they just opened it and read it before walking into class. But is that really reading? Active reading should be an engaging process if it is to be significant, if the reader is going to be affected or changed, even temporarily, in some way. Malcolm X makes this point in this piece. He became a literate, educated person who could communicate with others primarily because he took the time to study the words of his language. And he did it slowly and carefully.

In this slow process, the words entered into him. They got under his skin. They became his tools and his passion. Why? Because he did what those who teach study skills teach: One must read, write, speak, and reflect if one is to learn and learn well. Of course, in the end, he does throw in a twist: "Where else but in prison could I have attacked my ignorance by being able to study intensely sometimes as much as fifteen hours a day?"

Is this the key, though? To have the time? Or is there a way of making time? Another question: What is it that motivates him? Is it simply that he has time on his hands, or is it something else? What is his purpose in learning? Why does he want to appear smart?

Students have plenty of choices and ways of connecting with this piece through the discussion questions and through writing on their own experiences. It is always possible, too, to have students do a traditional rhetorical exercise by imitating, word for word, particularly intriguing sentences or passages. They can start out with one sentence, copying it for five days in a row, and then they can move to a paragraph. Instructors can choose one from the text that they find particularly fine, or they can find a favorite of theirs outside of the text.

The strategy here, of course, is the same that Malcolm X employed: to copy words, speak them, and think about them until they get under the skin. It might be interesting, too, to ask students to write about this experience. What, if anything, did they learn from it? Of course, to fully appreciate this way of learning, students may have to do this copying over the course of the semester. Even then, they may not appreciate the extent to which the words have taken over. As Malcolm X discovered, it takes time. Still, the seeds will have been planted.

Grace Paley, The Loudest Voice (text p. 211)

Suggested Cross-Reference Readings   *Text Pages*

Parish Streets	85
The Woman Who Makes Swell Doughnuts	68
The Son from America	95
Like Mexicans	113

We all have hidden talents, and in their journals students have a chance to "brag" about something they can do really well. This talent doesn't have to be artistic or educational. One student wrote about a "talent" she has for ignoring negative comments. Another wrote about a talent she has for sizing up strangers based on their wardrobes! Still another wrote about a talent he has for finding his way out of the woods. In other words, every student should be able to find something to be proud of.

In addition to boasting, students can also write an imaginative pre-reading piece in which they cast themselves in a starring role of their choice. Thus, students will be in the acting-showoff mode as they begin reading Grace Paley's delightful story about a young actress!

From the opening paragraph, readers should be able to draw one inference: The speaker in this piece has an honest, refreshing sense of humor as she looks back at her childhood. In addition, Shirley Abramowitz is proud to be herself, proud to be different, proud to be unique. She's pleased with herself and with her God-given talents (suggestion 1).

While Shirley is the central character in this piece, the tension swirls around her as her parents take sides on whether their daughter, a Jewish girl, should be in a traditional Christmas play. Some students may see the father as a conformist, a compromiser. He advises Shirley to be in the play so she can learn about the Christian culture. Shirley's mother, on the other hand, doesn't want her daughter to conform to the demands of the school. She (along with Mrs. Klieg) sees a direct conflict between Jewish beliefs and the Judeo-Christian approach to the season of Christmas, and thus would rather her daughter bow out of the play (suggestions 2 and 5). (If you teach students who are likely to be unfamiliar with Judaism, you'll need to discuss why being in a Christmas play disturbs Shirley's mother, the Rabbi's wife, and Mrs. Klieg. Many students may be unaware that Jews do not believe Jesus is the Messiah; lacking such basic understanding, Gentile students often fail to understand the central conflict in the story.)

Although Mrs. Klieg and Mrs. Abramowitz agree, Mrs. Abramowitz allows Shirley to be in the play, while Mrs. Klieg will not let her son participate. When it comes to such school matters, many students will agree that most parents tend to go along with what the school wants. Few parents feel comfortable making waves when it comes to public school policy, although students may sense that this trend is changing (suggestion 6).

Students will be divided about which parent's side they sympathize with more. Both parents present good points. The father sees this as a learning experience; the mother sees it as a disregard for Shirley's culture. You can promote further discussion on this issue by asking students to think of situations that parallel Shirley's dilemma. They may suggest controversies such as prayer in public places or flag salutations in school. In many towns across America, the Christmas crèche in a public place is also a point of controversy (suggestion 5).

To the Jewish people in this story, the Christmas tree on the corner, which had been decorated with public funds, was a symbol of disrespect for and ignorance of another culture. However, most of the Jewish people kept their feelings to themselves.

Likewise, even though Mrs. Abramowitz is against her daughter's performing in this play, she comes because she loves her daughter and wants to support her in the performance. Shirley is proud of her part in the play, and being the kind of mother that she is, Mrs. Abramowitz doesn't want the experience to be disheartening in any way for her daughter. One has the sense that the mother knows this disheartenment will come soon enough—when Shirley gets a bit older (suggestion 8).

Mrs. Abramowitz is not only a good mother, she is an intelligent woman, for she sees the irony in the fact that Christian children do not have the main roles in a play that reflects their basic beliefs. Clearly Mr. Hilton, the play's director, is concerned only with the "looks" of his play, not with the feelings or beliefs of his young students. His attitude seems to reinforce what Mrs. Abramowitz feels about Christmas in general: It is a grand show based on glittery performances with little basis in faith (suggestion 9).

Shirley's impressions may not be highly intellectual, but in her own way she also sees that this celebration, as symbolized by the Christmas tree, is a "lonesome" experience, for the Christians that she comes in contact with seem so joyless, so disunited from their beliefs and their cultures (suggestion 10).

If students want a change from the essay form, you might encourage them to rewrite any part of this story from another character's point of view. Such an approach would also reinforce the importance of perspective when it comes to viewing any problem.

If students feel strongly about the major conflict in this piece, encourage them to write an argument in which they state the major reasons for their position and their objections to the alternative.

Finally, students can write an analysis in which they assume their audience does not know the details of a particular cultural observance and they aim to explain, with appropriate details, the essence of this special occasion.

HEALTH: MIND AND BODY

Photographs

1. How do the images in the first photograph, including the images glimpsed in the mirror, suggest the connection between healthy bodies and healthy minds?

2. Create a brief scenario, with dialogue, description of setting, and stage directions to indicate what is happening in the second photograph. As you write, consider as many details as you can, including, for instance, the time indicated by the clock on the wall.

Paul Aronowitz, A Brother's Dreams (text p. 211)

Suggested Cross-Reference Readings	*Text Pages*
The Promised Land	62
Letter to My Mother	106
What Is Intelligence, Anyway?	176

As soon as we leave home, it seems, we better appreciate the nature of relationships and the people in our family. The awkwardness of adolescence and our dreams and disabilities are three topics students can think or write about before they begin Paul Aronowitz's piece in which he writes about similar issues.

Although this piece starts off rather detached—an explanation, really, of a brother's illness—as the piece progresses, it gets more personal. Most students will see the author as a kind man, a person whose feelings and reactions, both in the past and now, are understandable and commendable (suggestion 1).

When Aronowitz takes the reader back to the time when his brother's disease was yet to be diagnosed, most students would probably agree that in the same situation they, too, would be resentful. Mental illness is a disease that is often misunderstood. It is a common belief that a person should be responsible for his or her actions; many people refuse to acknowledge that mental illness can prevent sufferers from controlling their emotions as well as their actions. Mental illness is an invisible disease, one that even the medical community has long ignored (suggestion 2).

Schizophrenia is one type of mental illness. Through the portrait of Paul's brother, students will at least be able to see that people suffering from schizophrenia have a distorted view of reality. They often hear voices and see the world in a way that would not make sense to those of us who are mentally stable (suggestion 3).

In order to help students and herself understand the nature of schizophrenia, one instructor who used this piece with her class invited a spokesperson from the National Alliance of Mental Illness to visit her class. Since most areas have such speakers available, consider having an authority come to class to answer any questions or present further information on the topic of mental illness. The students will learn that Paul's brother represents millions of people in this country.

To move students to analyze the text closely, suggestion 5 asks them to look for one paragraph or line that they particularly liked. Consider having students model or paraphrase these paragraphs and come to some conclusions as to why they found these selected passages especially effective.

Once students have examined and discussed the topic of mental illness, you might want to discuss in what ways we can relate to people with this disease. The feelings of being capable or limited, the feelings and realities of never being able to reach a dream, are all possibilities for an extended writing assignment (suggestion 6).

In addition, it might be worthwhile to explore why some people reach their dreams while so many others do not. Do outside forces destroy the dream? Something else? If students also read any of the pieces recommended to accompany this one, they could also compare the circumstances, actions, and reactions described by those writers.

Florida Scott-Maxwell, Going Home (text p. 222)

Suggested Cross-Reference Readings   *Text Pages*

From Father, with Love	109
Mute in an English-Only World	151
Modern Courtesy	310

This is a perfectly beautiful piece, and it is one that is not ordinarily found in college anthologies. In fact, the author probably never imagined this piece as an essay in a writing text, as it is a part of her journal as she comes to terms with what it feels like to be old. The whole journal is exquisite, and this small piece is a tiny window into what few of us ever like to think about: growing old, being old.

The style is unpretentious, and the words are honest. Most students will have no problem connecting with this piece even though they have not been in Maxwell's predicament of being old and confined to a nursing home. What they can connect to is this: that human dignity fights for its life when one feels ill or helpless. They can also connect with older people they know who are facing the trials of old age—physical ailments as well as the prejudices people have against the very old.

However, this piece of writing is not at all sentimental or mushy. In fact, Maxwell avoids all whining or complaining in this piece. It is factual and honest, and she has maintained her dignity even in recounting the events of her helpless state. As they encounter the details of Maxwell's ordeal, readers will begin to see that a person should not be judged on his or her age any more than he or she should be judged on color, gender, nationality, or the like.

Certainly, then, this piece will elicit much discussion. In addition, instructors may want to have students compare the tone and style of this author with, say, a younger and more contemporary writer like Gary Soto ("Like Mexicans") or Meghan Daum ("Safe Sex Lies").

At the very least, students may notice how Maxwell, especially in the beginning, repeats specific words: *home, natural, pain, brave,* and *gift.* In fact, in some ways, this piece is more suggestive of poetry than prose because of its spare and simple style. What do students think of this? Which style do they prefer—this stark and bare wording or the more elaborate details of narratives such as Malcolm X's "Prison Studies"? Also, what more do they wish to see of her? Does the author leave anything out? One last thought for comparison: To whom is the author writing? Is it only to herself, or is it to someone else?

As a writing project, students could compose their own journal entry that is intended to be read by an outside audience. How would this writing differ from a personal diary entry? How would it differ from a formal essay? What does it accomplish that the other two cannot?

Eric Bigler, Give Us Jobs, Not Admiration (text p. 230)

Suggested Cross-Reference Readings	*Text Pages*
Like Mexicans	113
Unfair Game	282
Just Walk on By: A Black Man Ponders His Power to Alter Public Space	262

Many of the pieces in *Developing Connections* focus on perceptions and how people are influenced by what they see instead of what they know. To understand the position in which this author finds himself (handicapped and pitied instead of hired), students can think and write about their own difficulties in getting a job, the goals they have for themselves, and who they are—perspectives from both the outside and the inside.

Bigler uses a familiar structure in this piece, one that students might find helpful in their own writing. To make his point, he starts with a personal incident, threads the interview process throughout the piece, and then ends with the outcome of the personal incident. Such structure makes for a nice, tight essay because it introduces the audience to the main idea, repeats and reinforces the theme, and then wraps it up at the end by returning to the initial incident (suggestion 1).

Students may not be familiar with the phrase "Vocational Rehabilitation." However, after reading about the specifics of it, students will be able to see that this program is a way of getting people ready for a job, a vocation (suggestion 2).

Most readers would probably characterize the writer as a fair-minded, intelligent person. At this point, you might ask students their reactions when they meet people who use wheelchairs. Do they walk on the other side of the street? Duck into another store in the mall? Look the other way? Stare? Say to themselves, "Thank God it's not me" (suggestion 3)? If they think honestly about their own reactions, perhaps they'll be better able to understand why someone might be reluctant to hire a handicapped worker. What concerns would many employers have (sick days, ability to maneuver, customer reactions, etc.)?

One instructor who used this piece with her class asked a student who uses a wheelchair to come to the class and share her experiences with the students. To break down the barriers that often exist when a physical disability is apparent, the instructor did not tell the students ahead of time that the guest had a disability. Instead, she just asked students to write three or four questions that might help them to get to know a complete stranger. Questions ranged from general ("Where do you live?") to personal interest ("What's your favorite season?").

Bigler wrote this article to show others the difficulties that he faced simply because he used a wheelchair. Sympathy is fine, Bigler might say, but independence and respect are what people with disabilities, like nearly everyone, need most (suggestion 6).

Once students have discussed and thought about the issues that Bigler raises, you might direct them back to the pre-reading response in which they listed their future goals. Now they should imagine that they must use a wheelchair. In what ways are those goals now limited? Can any of them be achieved? What changes would be required for the student to pursue those goals?

Students are encouraged to write a definition paper on such abstract terms as *right, responsibility,* and *privilege.* To develop any of these terms, they should refer to pieces they've already read from the text.

Anna Quindlen, The War on Drinks (text p. 234)

Suggested Cross-Reference Readings   *Text Pages*

The Son from America	95
The Woman Who Makes Swell Doughnuts	68
Mothers, Sons, and the Gangs	123

This essay will strike a chord with practically every student in the class. It will also engage students in great discussions and, most likely, heated arguments. Why? As Quindlen discusses alcohol, she makes an argument that it is a destructive force in our society. Students' opinions in response to this claim will vary widely, and exchanges may become quite heated. You might consider asking students to imagine a country (or a planet!) where there is no alcohol. Would it be a better place? Why? Why not?

What do students think about the prevalence of alcohol in our society? What do they think of it as a ritual part of many celebrations? Certainly it is ingrained in many of our lifestyles, but Quindlen raises these tough questions: Why is it so popular, and what are the consequences of the widespread use of this legal drug?

Quindlen is, of course, a fine fiction writer as well as an essayist. Not only does this piece provide students with current and controversial subject matter, but it also offers a fine model for good writing. You might ask students to look carefully at the introduction. It details a short scene that immediately grabs the reader's attention. You may want to point out that an anecdotal opening remains one of the most popular introductions for seasoned writers because it is so effective in gaining readers' interest. Students might try out their own anecdotal opening on a variety of current topics. You might limit them to five or six lines that provide a small, detailed story leading into their subject. Then, too, they might take a great risk as Quindlen does and sculpt a one-sentence paragraph.

Besides the opening that invites and holds the reader, students should also notice the many ways in which Quindlen develops this piece. In addition to the opening anecdote (which does not have to be true, only probable), Quindlen switches to the views of current experts in the field: the surgeon generals. Then she brings in advertisements, without naming any specifics (why doesn't she?).

From these sources, she moves on to specific instances (no longer anecdotal stories). She uses direct quotes from children themselves. For some reason, students often do not readily see that dialogue is a fine way of "showing, not telling." It also is a technique that, like the anecdotal introduction, keeps the reader engaged. In the end, most students will find this piece readily accessible as well as entertaining and informative—the perfect blend for a good essay.

Now the question remains: Is the essay convincing? What do students think now? Has this essay changed their thinking in any way? To gain an objective perspective on the topic, students can, of course, review ads, commercials, and television to see how alcohol is presented. They can also talk to people (perhaps even each other) about the problems they've seen firsthand that have been associated with alcohol abuse. Without a doubt, every student will have something to add to this discussion—and while many, one can predict, will not have positive and glowing reports about the effects of alcohol, there will certainly be those who will defend responsible drinking.

Richard Selzer, The Discus Thrower (text p. 238)

Suggested Cross-Reference Readings	*Text Pages*
Fitting In	81
From Father, with Love	109
Modern Courtesy	310

Many students will find this piece difficult to read. Yes, it is clear and accessible, but the subject matter is grave, and the author does not hide the details of death and dying from his readers. However disturbing this piece may be, though, it is thoughtfully written and will almost certainly elicit strong responses from the students who read it.

A question arises in the first two paragraphs: Who is this essay really about? The opening is about the doctor, the "spy." The next paragraph focuses on the patient. This paragraph, by the way, is full of the most evocative writing in the essay. The metaphors are compelling, and if you want students to appreciate their power, it is worth stopping here and finding the connections between the actual and the symbolic—the frosted eyes as a snowbound cottage, the man's torso as a bonsai.

After the reader is introduced to the doctor and the patient, the two meet. The setting is dramatic. Students might stop and think of the implications of what is not in the room Likewise, they should pause to consider the dialogue, again, in order to draw conclusions as to the state of the patient's mind as well as the doctor's.

The next paragraph continues the conflict between the healer and the dying. Again, students might think about what's going on here. What is the doctor's tone? Why don't we hear any more from the patient now? We hear only one request: for a pair of shoes. What does this say about the patient? From what we've heard and seen thus far, what can students gather about who he is and what he is like?

The relationship with the nurses and aides takes on another dimension. It is not like that with the doctor. Students will want to reflect on these interactions and ask themselves if they would do anything differently. Could they do anything differently? Is the doctor a callous person? Is the patient an evil man? Are the nurses cold and heartless? At the bottom of it all, perhaps students can take a step back and think about the aura of death. What does it do to people? This, above all else, is the deepest question. It is the one that remains unanswered—unless, of course, the reader steps in and provides possibilities.

In many ways, this piece works as a short story, with a central character, a plot, a conflict, and a resolution. It is also rich in symbolism. If you choose, you might point out these elements that can appear in both fiction and nonfiction genres. Students may also compare this piece to Florida Scott Maxwell's "Going Home." In what ways are these selections similar? How are they different? By using these two pieces together, you can show students how to set up the "no-fail" comparison/contrast thesis by using the "although" starter. The students' focus, of course, is found in the independent clause following the dependent one. In addition, this structure points to a piece that centers on comparing or contrasting, as in the following examples:

> Although both pieces deal with the trauma that comes with physical pain and limitations, Maxwell's journal pictures hope and recovery, while Seltzer's essay shows despair and ultimate loss (contrast).

> Although Maxwell's journal deals with recovery while Seltzer's essay deals with loss, both writings point to hospital interactions that either leave a patient with hope or with a complete loss of self (comparison).

Jacqueline Navarra Rhoads, Nurses in Vietnam (text p. 242)

Suggested Cross-Reference Readings	*Text Pages*
Midterm Break	134
Safe Sex and White Lies in the Time of AIDS	320

Rich in action, details, drama, and meaning, this selection is the longest piece in the text. Assigning this highly charged reading will provide students with the opportunity to encounter a sustained narrative and thus will lead them toward the longer reading assignments they will receive in many of their college courses. Since college students are expected to read pieces of substance, learning to maintain attention is a necessary skill, and therefore you might encourage students to go along on this journey with open eyes.

Again, like many of the pieces in this section, "Nurses in Vietnam" presents gruesome details and situations that are disheartening. Because many students today have little in-depth exposure to the Vietnam War, you might want to provide a few background details. This piece takes them inside the war and will affect them on many levels—intellectually, emotionally, philosophically.

Readers should begin to get into the author's mind: What is Rhoads's purpose in writing this piece? Who is her intended audience? How would one describe the tone? For example, in paragraph 8 she writes, 'One guy had his face blown away, with hundreds of maggots eating away where his face used to be. Another guy, he had his eyes wide open." Why does the author use such colloquial language, calling these soldiers "guys"? What does she achieve by doing this? What other ways could she have approached this topic? Would any other ways have been as effective?

These types of questions push students to see that writers, indeed, have choices. Too often, students look for formulas in writing, and in some cases they have been provided with formulas (such as the five-paragraph essay) during their secondary school years. Perhaps learning to look beyond formulas is one of the main purposes for college writing courses. Students deserve to be made aware of how many options writers have and why these choices make a difference. In this piece, the author decides to simply stay herself—as one of the crew in Vietnam. There is a closeness here that is seldom seen in other circumstances. In fact, can students think of any situations that compare to this? What is it that the nurses do for one another? How do they sustain their energy?

Of course, there is a point in this piece that should cause all students to pause. In paragraph 17, the author, at a low point in this experience, asks the chaplain, "What are we doing? For what purpose are we here?" In context, the author is questioning the rationale behind the war. However, on a philosophical level, there are no two more important questions for students to ask of themselves. It may be an interesting in-class writing assignment to ask students, before the discussion begins, to attempt to answer these questions. What is important here, one understands, is not the answers per se. What matters is that the questions are asked.

This piece also raises questions that students might be anxious to explore: What takes away one's innocence? When was the last time you felt "young"? In paragraph 27 the author points to the things that have taken away her youth: seeing things she could never imagine, doing things on her own, making life-and-death decisions. However, when one thinks about it, don't many people experience similar challenges in their lives? Ask students to list their own experiences under each one of these categories, thus giving themselves a host of possibilities for future writing.

Besides the personal connections that students will be able to make in their writing, this piece provides the opportunity for students to research this topic. The internet is jammed with sites that include the writings and insights of Vietnam veterans. Tim O'Brien's *The Things They Carried* is a fine novel that many students would find provocative and fascinating should they want to read further about this war. Finally, students can debate and argue the need for war. Unfortunately, no matter when this piece is assigned, chances are that there will be a war going on in the world. Students can research the causes and effects of the present war, wherever it is. After all, there are nurses working there, too.

William Carlos Williams, The Use of Force (text p. 254)

Like Kate Chopin's "the Storm," this short story starts out in a matter-of-fact way. And yet, in the second sentence, the conflict becomes apparent: the daughter is very sick. Not just sick or ill, but "very sick." And while students may initially read this piece and see that the conflict is one between an angry doctor and a stubborn little patient, many conflicts are woven within this short story. Students should think about all the conflicts at work in this short dramatic piece: doctor and patient, husband and wife, doctor and the wife, doctor and the husband, doctor and himself. Thus, when paired this way, one can see that the central character is the doctor. He is the speaker, the one in charge, the one with the most conflicts. It is his show, so to speak. So what happens? What goes wrong?

To figure out the possible conflict within the doctor, students need to pay careful attention to the ways in which a reader gets to know a character: through his actions, his words, what others say about him, his appearance, his possessions, his values and beliefs. Students, in fact, could get into small groups and discuss what the story reveals about the doctor in each of these categories. This short piece may point students in a direction they would not have taken had they not approached this piece in an inductive manner, that is, looking at the evidence in order to draw conclusions.

Such thinking is not the norm when it comes to most American students. Studies show that while students in other parts of the world tend to think inductively—they gather evidence and then draw conclusions—Americans in general are deductive thinkers, stating a thesis and then giving the evidence. Of course, one way is not better than the other. But for students, it is an invaluable thinking experience to suspend their judgment until they look at the evidence.

For example, on the surface, when asked what their first impressions are, many students may be drawn to the doctor's side. The little girl, they may offer, is indeed a brat. However, if they look to the evidence first—the doctor's words for the little girl as well as the parents—they may see him in a different light. Indeed, he sees the child as a "heifer," a "cat," and a "savage brat." Further, he says the parents are "contemptible" to him and that this is now a "battle." What insights do these words give students into the doctor's persona?

Students should notice, too, that all the information comes from the doctor's perspective. How does this point of view compare to that in "The Storm"? What difference does it make as far as what the reader is allowed— and not allowed—to see? Why does Williams choose this point of view?

While most students will probably not be writing their own fiction in their college composition classes, they should still be encouraged to experiment with "point of view." For example, they could take a piece they've written from a first-person perspective and turn it into a third-person report. Similarly, they might take a third-person report and try writing it from a firsthand perspective.

Often students coming into a college writing class assume they are not allowed to use "I" when writing an essay. However, if this is correct, what does one say about the authors in this text? That they are bad writers? As most instructors recognize, the use of "I" is neither right nor wrong. It is a matter of purpose and audience. What does any given perspective do for the reader? What does it allow the writer to achieve? Some may say that these are sophisticated matters that most developmental writers do not need to bother with. However, many writers believe that perspective and audience are the most basic starting points of any piece of writing.

Besides personal reactions and conclusions that students draw from reading this piece, they can also debate the following: Is the doctor's use of force justified? In order to argue this point, students should use only the evidence of the story. In fact, it might make for an interesting class exercise to divide the students into assigned groups and have them argue back and forth, in an orderly, logical manner. That, of course, is the goal. What may happen, one can predict, is what happened to the doctor in this piece. While reason was the starting point, passion prevailed in the end. That's all right. It's even better if students come to recognize that they, too, may sometimes let emotion get the better of their reasonable selves. So much for judging the doctor too harshly.

MEN AND WOMEN

Photographs

1. What is the story behind the presentation of flowers in the first photograph? Who are the three people, and how are they connected to each other?

2. Describe any details of the second photograph that seem particularly striking. Then write a response to the photograph, focusing on your ideas about the changing roles of men and women.

Brent Staples, Just Walk on By: A Black Man Ponders His Power to Alter Public Space (text p. 262)

Suggested Cross-Reference Readings	Text Pages
Mute in an English-Only World	151
What Is Intelligence, Anyway?	176
Give Us Jobs, Not Admiration	230

While writing and talking about the pre-reading suggestions, most students will recall their strongly negative, usually angry responses to being misjudged, yet they also acknowledge that they, too, would be uncomfortable, even scared, if a stranger was walking close behind them at night. Thus, a common conflict arises: No one likes to be misjudged, and yet we all have a tendency to judge others mainly on their personal appearances. Why does this happen? Why is it so hard to trust, and why is fear such a natural reaction for us to have with one another?

When students read Staples's account, they will likely be divided about whether the young woman's reaction was justified. Have students point to the facts: It's late; the street is deserted; the woman is young and alone; the man is young, tall, casually dressed, and, as Staples points out, black. Ask students to think objectively here. If everything else were the same except it was daytime, would the woman be as scared? If the black man were wearing a suit and tie and were clean-shaven, would the woman be as afraid? Would the woman be afraid if it were night and the stranger walking behind her were a tall, casually dressed white man? In other words, Staples's point is that the woman is scared because he is black—does the example he gives support his point?

What students may see as they discuss and examine this piece further is that Staples is not blaming the young white woman here. As a matter of fact, he admits that women are vulnerable and "young, black males are overly represented among perpetrators of violence." His point is one that goes back to the students' initial pre-reading responses: No one should be judged because of the way he or she looks or on the actions of other members of their culture group (suggestion 4).

Overall, Staples addresses the problem of cultural bias related to males and, particularly, to black males. Toughness, he says, is expected. Students may or may not agree with his point here.

During his early years Staples avoided hanging out with the gangs. Many students will recognize that his education played a key role in keeping him focused on productive rather than destructive behavior. In the end, what Staples has discovered is how important appearance and demeanor are in the company of strangers. Staples's solution—whistling classical music while he walks late at night—reveals that he is not only a realist but also a creative person. Since people judge surface impressions, he will project a calm, intellectual air with his classical music. Through this quietly creative act, he may even alter some people's stereotyped views that all black men are dangerous and ought to be feared.

Staples's analogy of the cowbell reflects on the situation in which a black man often finds himself. A cat may be "belled" to warn birds (potential victims) of his approach. In this case, Staples "bells" himself, not to warn victims but to let people know that they are *not*, in fact, being stalked by a dangerous predator (suggestion 6).

Students can look further into the stereotype Staples describes by watching and categorizing the roles that blacks take on television shows or commercials. Again, this activity might lend itself to group collaboration and an audio-visual project. The more shows the group members view, the stronger inferences they can make.

Jon Katz, How Boys Become Men (text p. 268)

Suggested Cross-Reference Readings	*Text Pages*
Mothers, Sons, and the Gangs	123
I Show a Child What Is Possible	170
From Father, with Love	109

"I know lots of men who had happy childhoods, but none who have happy memories of the way other boys treated them," quotes Katz in this essay. From this statement alone, students should have plenty to say, plenty to think about, and plenty to write about. For starters, students can certainly draw on their own childhoods, specifically the years between fourth and eighth grade. In this time period, Katz claims, boys learn to play by the rules that make one a "man": act tough, don't cry, don't tattle, don't feel sorry for anyone, don't care—or at least don't act like you do.

Most students will easily connect to this piece, perhaps through a short, in-class writing. How did other boys or girls treat them in school? Most students will have no trouble remembering those school-day events—both good and not so good. In addition to their own personal connections, students should be able to look at this piece and question its truth: On what does the author base his assertions about how boys become men? If students take the time to look at this piece objectively, applying critical thinking and reading skills, they will see that all of Katz's claims are based around one incident in his childhood with Barry the bully, and one observation on a playground. The introductory scene may or may not be factual. Given that it is, then Katz's thesis—that boys are mean to each other and, therefore, learn to be insensitive grownups—is based on three examples.

The leap for students to take is this: Is this enough evidence on which to base such a claim? Of course, one may say that Katz is not writing a formal argument here. Indeed, he is not. His audience is women who read *Glamour* magazine. So for the audience, the piece may work.

But students can be shown another level of purpose in writing—and that is to give support and "proof" for what one writes. It is an exercise in thinking, really, to be able to do that. Therefore, instructors can ask students to find the statements in this piece that come across as "truth." With each statement they find, they should seek to find the support that backs up this truth. As critical readers, they may find that this piece is full of declaratives but is lacking evidence.

Of course, this piece can be taught, too, as a piece that is geared most of all to entertain. Elements that students can focus on include the introduction, the personal anecdote, the creative listing, the colloquial and "reader-friendly" language. Does this piece achieve its purpose? Probably. Is it entertaining? More than likely. Is it a solid argument that is developed with solid evidence and support? Not really. Could it be developed into a convincing argument? Perhaps.

If you would like to show how this piece can be brought to a level of convincing argument, have students brainstorm for areas where they could insert evidence. What would they need to find? Who else could they talk to? What facts and details would make this piece not only entertaining but credible? As some of the writing suggestions ask, students can make their observations, do their own interviews, and use their own experiences to come to their own conclusions: Do boys become men because of the way other boys treat them in childhood? From all of these perspectives, combined with the Katz essay, students will emerge with divergent but lively points of view.

Judith Ortiz Cofer, The Myth of the Latin Woman: I Just Met a Girl Named Maria (text p. 272)

Suggested Cross-Reference Readings	Text Pages
The Education of Berenice Belizaire	72
Like Mexicans	113
Mute in an English-Only World	151

There is no mystery as to what Cofer wants her audience to hear in this piece. In the second paragraph she writes, "I resented the stereotype that my Hispanic appearance called forth from many people I met." Her theme, therefore, closely mirrors Staples's point in "Just Walk on By: A Black Man Ponders His Power to Alter Public Space": Too often Americans judge others on superficial appearances, and these judgments are unfair and frequently humiliating.

Like Staples, whose piece appears in this same section, Cofer relies on personal examples that go through her childhood and into high school. Her experiences are rich in detail, and paragraph 5 is a magnificent example of "deep" development, where a writer slows down the action and zooms in with a telescopic lens. Many beginning writes tend to go much too quickly when they're explaining a moment in time. For example, they could write, "On career day I dressed how I thought I should to look professional and really was an outcast among my friends, still wearing their jeans and casual shirts."

A good exercise is to have students try this "slow down" themselves by writing one general statement of fact based on one event. ("The first day of school was horrible." "Football practice made me hate football forever." "I thought waitressing was going to be the most glamorous job in the world.") Once they have written this statement, they then have to come up with ten separate sentences that support and detail their opening.

In addition to using her own experience, Cofer brings in the experiences of other Puerto Rican women. She also sprinkles in examples of literature, the arts, and the media. Because she does this, her essay takes on a depth that allows students to see the synthesis of perspectives that makes for a fuller and more viable argument. Perhaps students could come to this conclusion on their own if they were asked to compare and contrast this piece with Katz's or Staples's piece. While the other two are certainly worthy pieces, three is no doubt that Cofer's piece is a fine model for full development and support of a thesis—an area that continues to be a weakness in most student writing.

Besides the example of depth and development that this piece provides, it also raises students' awareness as to how pervasive stereotypes of other cultures—Hispanics specifically—can be. Students may also realize, by watching television shows, how the media supports certain stereotypes. Then again, they may feel that the media's portrayal is fair. Until they actually observe a variety of television shows that focus on Hispanic characters, though, they will not be able to find this truth for themselves.

In the end, students could go back to any of their pre-reading responses and, based on what they have read and discussed, add or revise these initial writings to connect with Cofer's points. Thus, a pre-writing entry can grow into an extended essay, and that, too, is a valuable process lesson for writers to discover.

Charles Osgood, "Real" Men and Women (text p. 279)

Suggested Cross-Reference Readings　　　*Text Pages*

Marriage as a Restricted Club	118
The Son from America	95
"I Just Wanna Be Average"	193

If you want to be a bit sneaky, you might ask students to write a two-page draft on the topic that Osgood finds so offensive (both from his personal as well as his pedagogical viewpoint). Thus, they would already have written the first pre-reading question. Then you could discuss the problems they had writing this assignment. Using this approach, students may discover that they had the same difficulties as the student Osgood describes in this piece and that they came to the same conclusions that Osgood did!

Basically, Osgood believes that categorizing something as "real," is unfair and is, in many ways, a biased position. Therefore, he's approaching the assignment by contending that all people are "real"—unique, individual—and therefore there is no way to provide a definition that categorically includes some and rejects others (suggestions 1 and 2).

Osgood's approach begins by defining *real* as "alive." Students may also substitute words such as *actual* and *genuine*. These words work in Osgood's argument as well. For example, the opposite of *genuine* is *phony,* and therefore a "real" man is any one who doesn't pretend to be otherwise (suggestion 2).

When Osgood lists derogatory terms, he is trying to make the point that terms like *real* work like these words: They rob people of their identities and banish them into groups and camps. The words are offensive; they hurt and diminish human beings (suggestion 3).

If students have written their own responses to "real" men and "real" women, they should have little trouble comparing their views with

Osgood's. Initially, some students may have listed traits that characterize commonly accepted views of "real" men and "real" women. They may have touched on the stereotypical traits: Men are strong protectors; women are loyal and nurturing. Anyone who had this response should not feel like a failure here. Explain that the assignment was set up to elicit exactly that response, and that is why, in his last line, Osgood lambastes the English instructor (suggestion 5).

Susan Jacoby, Unfair Game (text p. 282)

Suggested Cross-Reference Readings	*Text Pages*
Finishing School	179
Nurses in Vietnam	242
Modern Courtesy	310

This piece might be a great one to pair up with Sanders's essay, "The Men We Carry," for in her essay, Jacoby shows the woman's point of view and suggests how she comes up with the image of men who hold the upper hand in the game of men and women in public settings. There are several points on which to contrast these two essays, including speakers, conflicts, and settings.

Jacoby's piece stands by itself as well, and this noted essayist's voice commands attention from readers in this piece. Students, too, will most certainly be divided in their reactions to Jacoby. Some will stand by her, while others, like the stranger at the bar in the opening scene, will find her rude and offensive. Students, of course, should be able to voice differing opinions. However, it is essential to ask them to point to places in the text that support their points. What does she say that proves she is rude? What does she say that gains your support?

Perhaps more than anything else, students may find out quite a bit about themselves in this piece by examining their reactions. Are they impressed by outspoken women? Are they offended by male strangers who flirt with women? What would they be apt to do in the same situation—and for what reasons would they act this way?

Ask students to evaluate the metaphor Jacoby uses about the social scene being a "game" with women as the ultimate "prize." Do students agree that today's social circumstances fit this metaphor? If so, does it annoy or bother them? How do male students' reactions to this piece compare to female students'? This might be a worthwhile survey to conduct in class. What conclusions can the students draw from their classmates' responses to this piece?

Students might also benefit from going through this piece and annotating the parts they agree with and those they disagree with. They could also discuss different ways the author could have handled the situation. In fact, it might be a good experience for the whole class to do some role playing to see how male and female students might handle the situations in which Jacoby found herself.

Perhaps more than anything, this piece will help raise students' awareness of the double standard that many believe is still apparent in this country. Extended writing activities encourage students to write their own "complaints" as well as their own codes of manners that might alleviate the tension women like Jacoby feel when they are out on their own. Students may also appreciate observing male/female encounters in order to see for themselves if what Jacoby describes is, indeed, a real part of the social scene.

Scott Russell Sanders, The Men We Carry in Our Minds . . . and How They Differ from the Real Lives of Most Men (text p. 286)

Suggested Cross-Reference Readings	Text Pages
Fitting In	81
From Father, with Love	109
I Show a Child What Is Possible	170

Although the pieces in this section all connect under the heading "Men and Women," they could also be grouped together under the title "Stereotypes." One might conclude, then, that two groups who have a tendency not to understand each other are, predictably, men and women. For, although they both fit under the term *human,* the differences and conflicts have been given widespread attention in every possible venue.

Sanders's piece reiterates this conflict. "What?" he muses, "Women think men are powerful and have treated women like second-class subordinates? Not in my life they haven't," he concludes. This is his perspective on the conflict he sees many young women having about the men in their lives. Paragraphs 10 and 11 are beautifully detailed, and students will be drawn to the portraits he paints in them. Not only will they be drawn to the weary faces and aching bodies of the men whom Sanders knew in his childhood, but they will, too, come to question the stereotype that some women hold of all men.

At the heart of this piece, what Sanders really calls into question, is this: On what do we base our views of "the truth"? It is a powerful question, and one that students ought to bring up time and time again as they think and read and write in this class. Is truth based on our own experience, or is it something else?

To show how truth is so much greater than our own experience, consider asking students how people know something is true. If you put a labeled circle on the board for each area a student mentions, then something huge emerges: There are seven or eight circles, all holding a certain truth: personal experience, others' experiences, observations, media, arts, books, experts, intuition. Which one is most convincing? As students might see, it is the blending of these areas that creates the most believable truth.

That is the point Sanders is making, or at least the question he implies. When two experiences differ, who is right? Again, students will come closer to the truth if they blend ways of seeking truth instead of relying on one or the other. The extended writing suggestions ask students to look outside of themselves—to look at family trees, students on campus, and statistics. What "truths" do these three areas reveal about them? Students may opt to work in groups and report their findings to the whole class. After doing this research, thinking about their own situations, and reading this essay, now what do they think?

Kate Chopin, The Storm (text p. 291)

Suggested Cross-Reference Readings	*Text Pages*
Marriage as a Restricted Club	118
The Loudest Voice	211
Safe Sex and White Lies in the Time of AIDS	320

Although this story was written in the nineteenth century, it is still highly controversial and evokes strong and conflicting responses from readers. Set this piece in its time frame for students so that they can see how earth-shattering it probably was at the time it was written. In fact, it was so controversial that Chopin was unable to publish it, although she was a well-known and popular writer of short stories.

Students may get into a conversation about the morality of actions and decisions in this short story, which can provide fruitful ground for critical thinking. However, in addition to the moral questions, students can be directed to see other factors at work in this story. For example, how does the setting connect with the plot? Do the characters change? What is Calixta's relationship with her husband? How do we know? What is Alcée's relationship with his wife? How do students know? How do the children enter into this story? What is their significance (for surely, the author had the choice of not including children). Further, what might the author's message be about men and women—or about love and marriage?

Finally, the central metaphor at work in this piece—that of the storm—is worth dissecting. What happens when a storm comes? What is the atmosphere like beforehand? What is it like afterward? What are the potential dangers associated with a storm? What are the benefits? If students take time to think simply about the storm, they can apply their observations to the relationship in the story. From here, they may be able to draw credible conclusions as to the author's purpose in writing this piece.

On one last note, students may also enjoy writing their own extended metaphors. "Love Is a . . ." or "Marriage Is a . . ." or "Passion Is a . . ." Of course, at some point, the metaphor breaks down. If students are ready for the deepest question, instructors might ask them to ponder this one: In what way is this relationship not like a storm? That should keep students thinking for a while.

CHOICES, ACTIONS, AND THE FUTURE

Photographs

1. Considering the theme "Choices, Actions, and the Future," what reason do you think the people in the first photograph might have for assembling around the sign that says, "We do not fear our enemies more than we love our children"? What issues are suggested by the sign and by the makeup of the group of people surrounding it?

2. Where is the woman in the second photograph standing? What choices and actions are suggested by the pamphlets she faces?

Charles Krauthammer, Of Headless Mice . . . and Men (text p. 301)

Suggested Cross-Reference Readings	Text Pages
The Son from America	301
Letter to My Mother	106
What Is Intelligence, Anyway?	176

This is a fine piece for teaching students the benefits of close reading. Because Krauthammer is a skilled wordsmith, students should, after reading through this piece once, go back and find words that are unfamiliar to them, trying to discover meaning from context or, when necessary, looking up definitions in the dictionary.

In addition to vocabulary, students should be intrigued with the references Krauthammer uses in this piece. What do they know of *Brave New World*? What do they know of *Frankenstein*? What are those two stories about? Have any of the students read them? You might want to set some background here, encouraging students who have read these novels to provide information and then supplementing with whatever further details are needed.

From here, students can go to the word level, extracting words that are the most powerful to them. And this essay is full of them: *monsters, headless, plundering, panicked, god, farm,* and so on. Each word carries its own connotation, and although some students may find it tedious, many may actually be intrigued to discover how a piece takes on greater significance the more a reader goes back to it and thinks about the powerful language that is being so skillfully employed by the writer. Actually, the key to Krauthammer's point is emphasized and developed by the emotionally loaded words he chooses.

Of course, students should also be urged to envision the other side of this argument—which Krauthammer never does. What could they say in defense of cloning? A little research will be needed here. In addition, students could research the cloning of Dolly the sheep. Where did this happen? Who is responsible? What was the scientific community's reaction? What was the public reaction? What do they think of Clinton's proposal? What are its shortcomings in light of the information they have on Dolly?

All in all, this piece will affect students on many levels and in many ways. It is a finely structured, persuasive piece that could provide a fine springboard for students' own platforms.

Rene Sanchez, Surfing's Up and Grades Are Down (text p. 305)

Suggested Cross-Reference Readings	*Text Pages*
Crossing the Border	76
"I Just Wanna Be Average"	193
The War on Drinks	234

Computers were not a part of the mainstream college experience ten years ago. Today, not only are they a part of the experience, but few students can get by without them. In this essay, Sanchez points out that while the computer age has certainly sped up students' access to information, it has also slowed down their learning insofar as many of them are becoming "addicted" to the other offerings of computers, particularly the huge and pervasive Internet.

This is not only a topic most students will be able to relate to but also one at which current psychology is now taking a closer look: How addicting are computers? In what ways do computer addicts compare to other addicts, such as alcoholics, drug users, or gamblers? If they are urged to research the most current articles by experts in this field, students may be amazed at what is being said about the power of computers today and the allure of electronic relationships. In fact, students may want to discuss how all this came to be. Why would someone be drawn to a relationship on a screen instead of one that grows and develops through personal contact? Why are so many people today taken in by what they read in a message on their computer screen that has been sent by a stranger? Why would students neglect studying for computer games?

This article provides plenty to think about. What parts of this piece are most convincing? What resources does Sanchez rely on? How effective are these sources? The piece, originally published in the *Washington Post,* is readable and clear. Students should have no problem following the information. In addition, Sanchez, unlike Krauthammer in "Of Headless Mice . . . and Men," gives both sides to this issue. Students could easily list the points on one side and then list those on the other. From this exercise, which side weighs in heavier? Which side of the argument would they take? What additional evidence would they offer to make their argument convincing?

Lore Segal, Modern Courtesy (text p. 310)

Suggested Cross-Reference Readings   *Text Pages*

The Woman Who Makes Swell Doughnuts	68
Fitting In	81
From Father, with Love	109

One aspect of our culture that many people believe has changed in the last two decades is the insistence on good manners. Readers who were born before the mid-1960s can remember the days when it was common for table manners and polite treatment of older people to be pounded into children's heads. Today, though, Segal suggests, there seems to be little regard for social manners. Before students examine the essay itself, they might begin by addressing this question: Is it true that manners are not important today? Should manners be considered important?

Students should depend on their own experiences at home and in the community to answer this question. What rules did their parents insist on as far as social courtesies go? How might their parents be different from their grandparents? How do children act in school today? What have students seen and observed in the last few months?

From this discussion, students may or may not agree with Segal's claim—that manners are becoming a thing of the past. The big question, though, is this: Should they be? Segal's piece is accessible and flows in an easy, smooth manner. She does not preach to the reader. In fact, the only "preaching" the reader hears are the conversations Segal has with her son. It's a clever way to get a point across without offending most readers.

Students can also make some good connections between their pre-reading responses to this piece and post-reading suggestions. Actually, pre-reading question 1 may elicit an entire classroom discussion or argument if students are willing to share the words they used to describe today's teenagers. Besides sharing these responses, though, students can move one step higher in their thinking: So what do these words reveal about today's teenagers? And further, are these the same words you would use to describe teenagers in the past?

In the end, students can suspend their opinions and set out to observe teenagers and young college students in natural settings. How are the manners observed in these places? As another thought, some students in the class should also be directed to observe adults' manners in restaurants and stores. What do the observations reveal about the manners adults employ today? A comparison like this may, of course, undermine a bit of Segal's premise: It's not teens alone who have lost their manners today; it's Americans in general. Then again, observing two separate groups may serve to support Segal's own contentions.

Samuel Francis, Illegal Motives (text p. 313)

Suggested Cross-Reference Readings	*Text Pages*
Mute in an English-Only World	151
Words That Hurt, Words That Heal:	
How to Choose Words Wisely and Well	155
Prison Studies	206
Just Walk on By: A Black Man Ponders	
His Power to Alter Public Space	262

As you begin teaching this selection, you might consider asking students to define standard political terms such as *progressive, liberal, right wing, left wing,* and so on. Some students will almost certainly know these terms, and understanding the difference between liberal and conservative political philosophy will be a useful concept for all students.

While the article Francis discusses appeared in a British journal, the controversy focuses on current concerns within the United States. Students might want to consider whether the issue of hate crimes even comes up on the campus where they are studying. At many colleges, administrations respond much more harshly to students who harass other students when the motive appears to be prejudice and hate. For example, graffiti attacking someone's ethnic or religious background would be taken more seriously than messages making fun of a person's clothes or personal appearance, even though the language might include obscenities. What do students think of this? Should the motive be considered when the punishment is meted out?

To consider ways of building argument, you might have students look at the evidence or proposition in each paragraph and ask them to determine not only how convincing they find it and why but also how someone who held a view opposed to Francis's might refute each of his points.

If you choose to assign long-term projects related to this topic, students might look for articles in journals and newspapers discussing the issue of hate crimes or study Web sites that promote hateful ideas to determine how dangerous they find these sites. Should they be banned? Regulated? How does the concept of first amendment rights fit into this argument?

Linda Chavez, There's No Future in Lady Luck (text p. 317)

Suggested Cross-Reference Readings	*Text Pages*
Crossing the Border	76
The Son from America	95
A Brother's Dreams	221

In Segal's essay, the author argues that modern courtesies have gone by the wayside in America. In this piece, Linda Chavez argues that the American work ethic has been replaced by the notion that the luck of the draw is the way to success. Students will have no trouble grasping this piece, but the challenge here is to look closely at Chavez's support. For what arguments does she provide good support, and in what statements is she simply begging the question?

One way for students to become critical readers—that is, for them to question everything they read in order to discern if it is credible and fair—is to go through and separate fact from opinion. It's a great exercise to use with this piece, for Chavez takes a common track here. Some of what she writes is fact; some of what she writes is opinion stated as facts.

There are two reasons students need to know the difference. First, unless they become critical readers and listeners, they will be easily swayed in any direction by spurious arguments. Second, students can be helped in their writing and speaking by becoming aware of how often they may rely on opinion to support their arguments. Of course, there is a spectrum here. On one end is pure unfounded opinion. On the other end are well-used facts. In between fall skewed facts and lightly supported opinions.

Examples of how this piece falls on the continuum are as follows: Paragraph 4 is based on facts. The next paragraph is a declarative sentence supported by a quotation from one woman. It, then, is a lightly supported opinion. Then Chavez claims the lady is typical. There is some evidence to support this, as Chavez then includes a probable observation. But the opinion is still on shaky ground. In paragraph 12, though, Chavez writes that the lottery money comes from "the least educated most gullible" people in our society. She says it is "a fact," but is it? No support is given, except for the next paragraph, in which someone says the poor are lotteries' best customers. Then the poor are the least educated and most gullible? Is this fair? Is it a fact?

Through a careful study of the author's claims, students should be able to come to a fair conclusion: Is this piece believable and credible, or not? In general, Chavez develops a credible argument, but there are certainly places where some carefully researched data would provide additional and important support.

Meghan Daum, Safe Sex and White Lies in the Time of AIDS (text p. 320)

Suggested Cross-Reference Readings	*Text Pages*
Crossing the Border	76
Three Poems on Words	164
The Use of Force	254

The structure and argument in this essay may challenge students initially. At times, it seems as if the author is contradicting herself. The introduction is, without a doubt, an attention getter, and here it seems as if the author is a person with great fear. In fact, in this opening, she says she is "terrified of this disease"—this disease being AIDS.

The contrasting argument, written in a colloquial style, is all compactly presented in the second paragraph. According to Daum, we don't stick to the rule of precaution because we can't. We can't—why? "Because it's not fair." What isn't fair? The author doesn't yet say. Instead, she goes on to say "our sense of entitlement exceeds our sense of vulnerability." Sense of entitlement to what? Again, the author does not yet say.

Students may find themselves asking the questions suggested above as they complete this paragraph. "She said what?" they might ask. This whole paragraph runs on like a speeding commuter train, without stops, without conversation. So now students have to ask, Why does she write her argument like this? What could possibly be the reason?

Students may find that there are some good reasons for writing in this manner. One, of course, is to get readers to ask questions that will encourage them to read on. When they do, the argument becomes clearer. The language and style change; they slow down, and Daum creates a careful, logical argument. Students should try to find out what the point of each paragraph is. Suspending discussion of Daum's argument might be a good idea, since summarizing and understanding this piece are in order before one can agree with or attack her premises.

Students will find the author saying that conflicting messages have come from those in control—parents, teachers, government agencies. On one hand, they tell us sex is a good thing. On the other, they say, "Watch out. It can kill you." This author eventually concludes that the latter message has not been delivered in a truthful or effective manner.

This piece, while a bit complicated, will give students plenty to discuss. Some of their discussion might center around their impression of the author itself. They may also write and talk about the way that AIDS is presented today. Are the messages inconsistent? What do they think from their own education and experiences with the media's messages?

Ultimately, students ought to think about this issue on their own. Why does Daum write this essay? To convince them to have unprotected sex, or to look more closely at the forces—primarily schools, media, and the government—that shape and control their lives? One book that may be used along with this last theme of how many forces shape our decisions and our lives is Mary Pipher's *Reviving Ophelia*. Ironically, while Pipher agrees with Daum on the conflicting messages that target young people, most especially girls, Pipher might counter that Daum should think carefully before making what might become earth-shattering decisions.

Kurt Vonnegut, Harrison Bergeron (text p. 327)

Suggested Cross-Reference Readings	*Text Pages*
Zipped Lips	147
"I Just Wanna Be Average"	286
Discus Thrower	238

Many of the underlying messages of the pieces in this section address the issue of what controls people's actions and how people can be so easily controlled. While it is fiction and set in the future, "Harrison Bergeron" also reiterates this theme, thus providing a unifying idea for the section.

Whether it be media, computers, lotteries, or scientific advance, people's behaviors are, to a very real extent, controlled by outside forces. Vonnegut's piece shows an extreme control: that of a government controlling its citizens' thoughts. In fact, thinking and questioning, the ways to independence and freedom, are the very aspects the government aims to control.

The main question here, then, is this: Is this story completely far-fetched? Is it simply entertaining—albeit frightening—science fiction? If students have learned to raise questions, be curious, and find truth as well as flaws through their readings of the pieces in this text and their own writings, then they may well come to this conclusion: It is not as far-fetched as we would like it to be.

Certainly students can study the characters in this piece. What is George like? How about Hazel? Are they the same? How? How not? Most important, how did they get to be this way? Did they ever have any choice? Did they see it coming?

And then there's Harrison and Diana Moon. How do they compare to the parents? How do they differ? How are they like one another, and how do they differ? Perhaps the question most students should raise on their own is this: What causes Harrison to break free? What is Vonnegut suggesting to his readers?

Students might also think about how the acts of reading and writing differ from watching television or listening to headsets. What are the conveniences that actually control us? Students can write based on their personal experiences: "What one modern convenience do you depend upon most, and what would your life be like without it?"

The presence of technology and the way it controls humans is an intriguing topic. This story appears at the end of this text, but it is a great beginning to motivate readers to question the possibilities of excellence in their own lives, examining whether their talents have been encouraged or whether they have been urged to act more like everyone else, abandoning their unique gifts. The story might also raise the question of whether some gifts in our society are privileged over others and whether pride in individual accomplishment is overshadowed by exhortations to contribute to team or group efforts.